Yakov Perelman

Algebra for Fun

For general information on our products and services, please contact us on prodinnova@mail.com

Printed in the United States of America.

ISBN : 9782917260265

10 9 8 7 6 5 4 3 2 1

Algebra for Fun

Contents

Chapter One

THE FIFTH OPERATION OF MATH-EMATICS

The Fifth Operation

Algebra is often called the "arithmetic of seven operations", thus stressing the fact that the four commonly known mathematical operations are supplemented by three new ones: raising to a power and its two inverse operations.

We start our algebraic talks with the fifth operation -raising a number to a power.

Do the practical affairs of everyday life have any need for this operation? It turns out they do. We often encounter such situations. First recall the many cases of computing areas and volumes that ordinarily call for squaring and cubing figures. Then the force of universal gravitation, electrostatic and magnetic interactions, light and sound that diminish in force in proportion to the second power of the distance. The periods of revolution of the planets about the sun (and of satellites about planets) are connected with the distances from the centre of revolution by a power relationship: the squares of the periods of revolution are in the same ratio as the cubes of the distances.

One should not think that real life involves only second and third powers with higher powers found only in the problems of algebra books. Engineers dealing with strength of materials calculations constantly find themselves up against fourth powers and even (when computing the diameter of a steam pipeline) sixth powers. A water engineer studying the force with which running water entrains rocks also has to deal with a sixth-power relationship. If the rate of the current in one river is four times that in another, the fast river is capable of rolling rocks that are 4^6, or 4096, times heavier than those carried by the slower river.[1]

Even higher powers are encountered when we study the relationship between the brightness of an incandescent body (say the filament in an electric light bulb) and its temperature. The total brightness (luminance) in the case of white heat increases with the twelfth power of the temperature; in the case of red heat, it increases with the thirtieth power of the temperature (we refer here to absolute temperature, that is, we reckon from minus 273°C). This means that a body heated say from 2000° to 4000° absolute temperature, that is by a factor of two, becomes brighter by a factor of 2^{12}, which is more than 4000 times greater. We will come back to this remarkable rela-

-1- For more about these things see my book Recreational Mechanics (in Russian).

tionship when we deal with the manufacturing of electric light bulbs in a later chapter.

Astronomical Numbers

Hardly anyone makes as much use of the fifth mathematical operation as astronomers, who are constantly dealing with enormous numbers that consist of one or two significant digits and then a long string of zeros. If we were to write out in full all of these truly "astronomical" numbers, it would be very inconvenient, especially in calculations. Take the distance to the Andromeda Nebula. Written out in full in the ordinary way, we have the following number of kilometers:

95 000 000 000 000 000 000

Now if we were to express this distance in a smaller unit, say, centimeters -and this often happens in astronomical calculations- we would have to add five more zeros:

9 500 000 000 000 000 000 000 000

The masses of stars call for still bigger numbers, especially if they are expressed in grams (and many calculations call for grams). The mass of the sun expressed in grams comes to

1 983 000 000 000 000 000 000 000 000 000 000

It is easy to imagine the difficulties involved in calculating with such unwieldy numbers and also the chances of making mistakes. Yet the abovementioned numbers are by far not the largest to be found in astronomy.

The fifth mathematical operation serves as a simple way out of this complication. The number one followed by a string of zeros is always a definite power of ten:

$100 = 10^2$, $1000 = 10^3$, $10\ 000 = 10^4$ and so forth.

The two giant numbers given above can now be written down neatly as 95×10^{23} for the first number and 1983×10^{30} for the second number.

This is done not only to save space but also to simplify calculations. If we needed to multiply these two numbers together, all we would have to do is find the product $95 \times 1983 = 188\ 385$ and put it in front of the factor

$102^{23+30} = 10^{53}$:

$$95.10^{23} \times 1983.10^{30} = 188\ 385.10^{53}$$

This is of course much more convenient than to write out a number with 23 zeros and then one with 30 zeros and, finally, with 53 zeros. Not only is it more convenient but also more reliable, since it is easy to miss one or two zeros in such long strings of them, and the result would be wrong.

How Much does the Earth's Atmosphere Weigh?

Here is a convincing way to simplify practical calculations by means of exponential notation (using exponents to express powers of numbers): let us determine how many times the mass of the earth is greater than the mass of the earth's atmosphere.

As we know, each square centimeter of the earth's surface supports a column of air equal to one kilogram. The atmospheric shell of the earth is, as it were, made up entirely of such columns of air -as many as there are square centimeters on the earth's surface. That is how many kilograms the atmosphere of our planet weighs. From a reference book we find the earth's surface to be equal to 510 million or $51\ 10^7$, square kilometers.

Now figure out how many square centimeters there are in a square kilometer. A linear kilometer contains 1000 meters with 100 centimeters in each, which means it is equal to 10^5 cm, and a square kilometer contains $(10^5)^2 = 10^{10}$ square centimeters. And so the earth's surface works out to

$$51\ 10^7\ 10^{10} = 51\ 10^{17}$$

square centimeters. And that also is the weight of the earth's atmosphere in kilograms. Converting to (metric) tons, we get

$$51\ 10^{17} : 1000 = 51\ \ 10^{17} : 10^3 = 51 \cdot 10^{17-3} = 51.10^{14}$$

Now the mass of the earth is taken to be 6.10^{21} tons.

To determine how much heavier the globe is than its atmosphere, we perform the following division:

$$6 \cdot 10^{21} : 51\ \ 10^{14} \approx 10^6$$

which means that the mass of the atmosphere is roughly one millionth of that of the earth.[2]

Combustion Without Flames or Heat

Ask a chemist why wood or coal burns only at a high temperature and he will tell you that strictly speaking carbon combines with oxygen at any temperature, but that at low temperatures the process is extremely slow (which means a very small number of molecules enter into the reaction) and so is never detected. The law that governs the rate of chemical reactions states that a drop in the temperature of 10° reduces the rate of the reaction (the number of participating molecules) by a factor of two.

Let us now apply this to the reaction of wood combining with oxygen, which means "burning". Suppose at a flame temperature of 600°, one gram of wood burns up completely in one second. How long will it take for one gram of wood to burn up at 20°? We already know that at a temperature that is lower by 580 = 58 10 degrees, the rate of reaction is less by a factor of 2^{58} which means 1 gram of wood will burn up in 2^{58} seconds.

How many years is that? We can get an approximate answer without performing 57 multiplications by 2 and without using logarithmic tables. Let us make use of the fact that

$$2^{10} = 1024 \approx 10^3$$

Hence

$$2^{58} = 2^{60-2} = 2^{60} : 2^2 = (1/4)\ 2^{60} = (1/4)\ (2^{10})^6 \approx (1/4)\ 10^{18}$$

which is about a quarter of a quintillion seconds (quintillion as in the American and French system of numeration). There are 30 million (or 3.10^7) seconds in a year, and so

$$(1/4)\ 10^{18} : (3.10^7) = 1/12.\ 10^{11} \approx 10^{10}$$

Ten thousand million years! That is roughly how long it would take for a gram of wood to burn without flames and heat.

Thus, wood and coal burn at ordinary temperatures without even being set fire to. The discovery of tools for making fire has accelerated that impossibly slow process by a factor of thousands of millions.

-2- The symbol ≈ stands for "approximate equality".

The Changing Weather

PROBLEM

Let us describe the weather using only one characteristic: cloudy or not cloudy. Days will be described as clear or overcast. Do you think there will be many weeks with different changes of weather under this condition?

There would appear to be very few: a month or two will pass and all combinations of clear and overcast days in the week will have been exhausted. Then one of the earlier combinations will inevitably recur.

But let us calculate exactly how many distinct combinations are possible under these conditions. This is a problem that unexpectedly leads us to the fifth mathematical operation.

The problem is: in how many different ways can clear and overcast days alternate in one week?

SOLUTION

The first day of the week is either clear or overcast: that gives us two combinations.

Over a two-day period we have the following possible alternations of clear and overcast days:

clear and clear,

clear and overcast,

overcast and clear,

overcast and overcast.

Thus, in two days we have 2^2 distinct alternations. In a three-day period, each of the four combinations of the first two days combines with two combinations of the third day; there will be $2^2.2 = 2^3$ alternations in all.

In four days the number of alternations will reach $2^3.2 = 2^4$.

In five days, there will be 2^5 alternations, in six days 2^6, in seven days (one week), $2^7 = 128$ distinct alternations.

Hence, there are 128 weeks with a distinct sequence of clear and overcast days. A total of 128 .7 = 896 days will pass before one of the earlier combinations is repeated. A repetition can of course occur before that, but 896 days is the period after which a recurrence is inevitable. And conversely, two years and more (2 years and 166 days) may pass during which the weather in one week will not be like that of any other week.

A Combination Lock

PROBLEM

A safe was discovered in an old building, but no one knew the combination of letters out of the 36 on the rim of each of the five circles that would unlock the safe. So as not to break the safe, it was decided to try all combinations of the letters on the rim at the rate of three seconds for each new combination.

Was there any chance of hitting upon the right combination within the next 10 days?

SOLUTION

Our task now is to calculate the total number of combinations of letters that have to be tested.

Each of the 36 letters of the first circle can combine with each of 36 letters of the second circle. This brings the total of two-letter combinations to

$$36.36 = 36^2$$

To each of these combinations we can adjoin any combination of 36 letters of the third circle and so there are $36^2.36 = 36^3$ possible combinations of three letters.

Continuing in the same way, we obtain 36^4 possible four-letter combinations and 365 five-letter combinations, or 60 466 176. To run through these 60-some million combinations would require (at the rate of 3 seconds per combination) 3 x 60 466 176 = 181 398 528 seconds, or 50 000 hours, or nearly 6300 eight-hour working days, which is more than 20 years.

Thus, the chances that the safe will be opened within 10 working days, is 10 out of 6300 or one in 630. This is a very small probability.

The Superstitious Cyclist

PROBLEM

Cyclists have six-digit license numbers on their bicycles. A man bought a bicycle but, being superstitious about a possible crash-up, called "figure-eight" in cyclists' terminology, he was worried about an "8" appearing in his license number. However, after some thought, he calmed down realizing that 10 digits (0, 1, ..., 9) can take part in each number and there is only one unlucky digit, 8. Therefore there is only one chance in ten of his getting an unlucky number.

Was his reasoning correct?

SOLUTION

There are 999999 numbers in all: from 000001 to 999999. Now let us see how many lucky numbers there are. The first position can accommodate any one of the nine: lucky numbers: 0, 1, 2, 3, 4, 5, 6, 7, 9. The second position too. And therefore there are $9 \times 9 = 9^2$ lucky two-digit combinations. To each of these combinations we can adjoin (in the third position) any one of nine digits and so we come up with $9^2 \times 9 = 9^3$ lucky three-digit combinations.

In the same way we determine the number of six-digit combinations that are lucky: 9. True, one has to take into account the number 000000, which cannot be used for a license plate. Thus, the number of lucky license numbers is $9^6 - 1 = 531\,440$, which is slightly more than 53% of all the numbers and not 90%, as our cyclist had presumed.

We leave it to the reader to figure out that there are more unlucky numbers than lucky ones among seven-digit numbers.

The Results of Repeated Doubling

A striking example of an exceedingly fast build-up of some small quantity when repeatedly doubled is the famous legend about the award to be given to the discoverer of chess.[3] Here are some other examples, less famous.

-3- See my book *Figures for Entertainment.*

PROBLEM

The infusorian paramecium divides in half on the average every 27 hours. If all newly born infusorians remained alive, how long would it take for the progeny of one paramecium to fill up a volume equal to that of the sun?

Starting data: the 40th generation of a paramecium, when none perish, occupies one cubic meter; we take the volume of the sun as equal to 10^{27} cubic meters.

SOLUTION

The problem reduces to determining how many times 1 cubic meter has to be doubled in order to obtain a volume of 10^{27} cubic meters. Since $2^{10} \approx 1000$, we have

$$10^{27} = (10^3)^9 \approx (2^{10})^9 \approx 2^{90}$$

This means that the fortieth generation has to undergo another 90 divisions before it attains the volume of the sun. And the total number of generations, from the first on, comes out to $40 + 90 = 130$. It is easy to calculate that this will occur on the 147th day.

In passing, we may note that microbiologist Metalnikov actually observed 8061 divisions of paramecium. I leave it to the reader to figure out what a staggering volume the last generation would occupy if not a single infusorian had perished.

Another interesting problem is the converse of the one just proposed.

Imagine the sun to divide in half, then the halves also in half, and so on. How many divisions will it take to reduce the sun to the size of an infusorian?

Although the reader already knows the answer, 130, it seems strikingly out of all proportion.

This same problem was proposed to me in the following form.

A sheet of paper is torn in half, one of the halves is again torn in half, and so on. How many divisions will it take to reduce the paper to the size of an atom?

Assuming the paper weighs one gram and an atom is taken to be on the order of $1/10^{24}$ gram, we reason as follows. Since it is possible to replace 10^{24} by the approximately equal expression of 2^{80}, it is clear that only 80 halving operations will be required, which is nowhere near the millions that one often hears as an answer to the problem.

Millions of Times Faster

An electric device called a trigger (or flip-flop) circuit contains two electron tubes (transistors or so-called printed circuits can take the place of electron tubes). In a trigger circuit, current can flow through only one tube: either the left-hand one or the right-hand one. The trigger circuit has two contacts to which a transient electric signal (pulse) can be fed and two contacts through which the response from the circuit is delivered. When an electric signal is fed to the trigger, it switches: the tube that was conducting current is disconnected, and the current passes through the other tube. The response signal is delivered by the circuit when the right-hand tube is non-conducting and the left-hand tube is conducting.

Let us see how the trigger circuit will operate when several electric signals are delivered to it in succession. We will describe the state of the circuit by its right-hand tube: if no current flows through the right-hand tube, we say that the circuit is in "state 0", if current flows, then it is in "state 1".

To start, let us suppose the circuit is in state 0, which means the left-hand tube is conducting (Fig. 1). After the first pulse, current will flow through the right-hand tube and the circuit switches to state 1. In this state there is no response signal from the circuit since that can occur only when the right-hand (not the left-hand) tube is off.

After the second signal, the current flows through the left-hand tube, and the circuit is again in state 0. But in this state the trigger circuit delivers a response signal (pulse).

Then, after two pulses, the circuit returns to its original state. Therefore, after the third pulse, the circuit is in state 1, just as it was after the first pulse. After the fourth pulse the circuit is in state 0 (like it was after the second pulse) with a simultaneous delivery of the response signal, and so on. The state of the circuit is repeated after every two pulses.

Now suppose there are several trigger circuits and that signals are delivered to the first circuit, responses of the first circuit are delivered to the second circuit, responses of the second are delivered to the third, and so forth (in Fig. 2 the trigger circuits are arranged from right to left). Now let us see how such a chain of trigger circuits will operate.

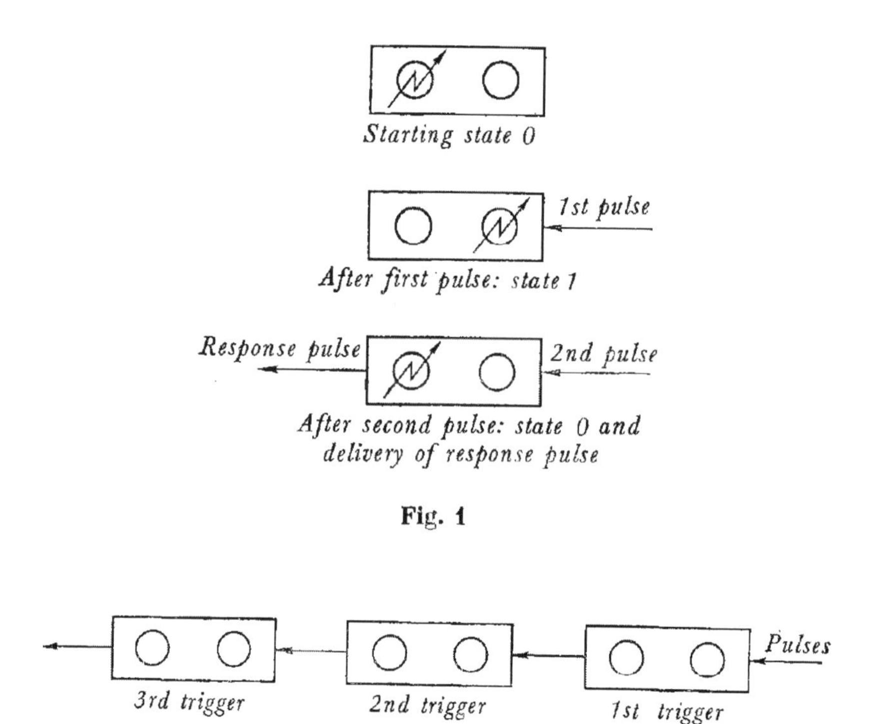

Fig. 1

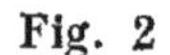

Fig. 2

To begin with, suppose all circuits are in state 0. For instance, for a chain consisting of five circuits we have the combination 00000. After the first signal, the first (right-most) circuit is in state 1, and since there is no response pulse, all other circuits are in state 0, and the chain may be described by the combination 00001. After the second signal, the first circuit becomes non-conducting (state 0), but will deliver a response signal that switches on the second trigger circuit. The other circuits remain in state 0, and we get the combination 00010. After the third signal, the first circuit goes on, and the other circuits remain in their states. We get the combination 00011. After the fourth signal, the first circuit goes off and delivers a response signal; the response turns off the second circuit and also yields a response; finally, this latter signal turns on the third circuit and we get the combination 00100.

Continuing in the same manner, we finally get the following set of combinations:

1st signal — combination 00001

2nd signal — combination 00010

3rd signal — combination 00011

4th signal — combination 00100

5th signal — combination 00101

6th signal — combination 00110

7th signal — combination 00111

8th signal — combination 01000

We see that the chain of trigger circuits "counts" the incoming signals and records them in a system of zeros and ones. It is easy to see that this system is not our familiar decimal system of numeration but the binary system.

In the binary system of numeration, every number consists of ones and zeros. The unit of the next higher order is two times that of the preceding order, and not ten times greater, as it is in the decimal system of notation. The unit in the rightmost position in binary notation is the ordinary unit. The unit in the next higher order (or place) -second from the right- denotes two, the next order denotes four, the next eight, and so forth.

For example, the number 19 = 16 +2 +1 is written in binary as 10011.

And so we have a chain of trigger circuits that counts the number of incoming signals and records them in the binary system of notation. An important thing to note is that the switching of a circuit (that is, the recording of one incoming pulse) takes a few hundred millionths of a second! Modern circuits can count tens of millions of pulses per second, which is millions of times faster than what a human being can do without any instruments: the human eye can distinguish the signals in a sequence if they don't come faster than one every tenth of a second.

If we make up a chain consisting of twenty trigger circuits, which amounts to recording the number of signals by means of at most twenty digits in binary notation, then we can count up to 220 — 1. This number exceeds a million. Now if we made up a chain of 64 circuits, we could write down the famous chess number.

The ability to count millions of signals a second is very important in experimental studies and nuclear physics. For example, it is possible to count the number of particles of a certain kind that fly out of a nucleus in radioactive disintegration.

Ten Thousand Operations per Second

It is a remarkable thing that trigger circuits permit one to perform operations on numbers. Let us examine the addition of two numbers.

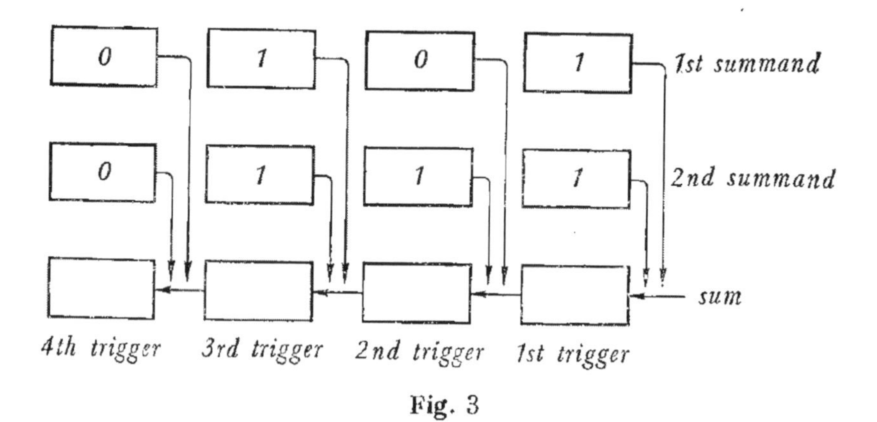

Fig. 3

Suppose we have three chains of circuits connected as shown in Fig. 3. The topmost chain serves to record the first term (or summand), the second chain records the second term, and the bottom chain gives the sum. When the device is switched on, the circuits in the lower chain receive signals from those in the upper and middle chains that are in position (or state) 1.

Suppose, as is shown in Figure 3, we have the summands 101 and 111 (in binary notation) in the first two chains. Then the first (rightmost) trigger circuit of the bottom chain receives two signals when the device goes on: these signals come from the first circuits of each of the summands. We already know that the first circuit remains in state 0 upon receiving two signals, but it sends a response to the second circuit. Besides, the second circuit receives a signal from the second summand. Thus, the second circuit receives two signals and thus is in state 0 and sends a response pulse (signal) to the third circuit. Besides that, the third circuit receives two more signals (from each of the summands). Upon receiving three signals, the third circuit goes to state 1 and sends a response signal. This response carries the fourth circuit to state 1 (no other signals are sent to the fourth circuit). As a result, the device shown in Fig. 3 performed, in binary notation, the addition of two numbers by columns:

101

+111

= 1100

or, in the decimal system of numeration, we have 5 + 7 = 12. The response pulses in the lower chain of circuits correspond to the carrying operation: the device "remembers" one unit and carries it over to the next order (column).

Now if each chain had, say, 20 circuits instead of 4, then we could add numbers within a million, and with more circuits the numbers could be increased still further.

Actually, of course, the add unit has a more complicated electrical network than that shown in Fig. 3. For instance, the device has special circuits that delay the signals. The point is that if signals from both summands arrived at the first trigger circuit of the bottom chain at the same time (that is, when the device is switched on), then they would merge and be received as a single signal. This is circumvented by having the signals from the summands delayed so that they come in one after another. As a result the addition of two numbers takes more time than the more recording of a single signal in a flip-flop counter.

By changing the circuit system we can make the device subtract. Multiplication is also possible (it amounts to successive addition and therefore takes several times as long as addition) and division and other operations too.

The devices we have just described are used in modern computers, which are capable of hundreds of thousands, even millions of operations per second! Why millions, you might ask. Is it really necessary? For instance, what difference would it make if a computer spent one ten-thousandth of a second (even a quarter of a second) more time in squaring a 15-digit number? Aren't both practically instantaneous?

Don't hurry with your conclusions. Let us take a simple case. A chess master makes a move only after analyzing tens, sometimes hundreds of possibilities. Now, if studying a single variant takes several seconds, then a mental investigation of hundreds of possible moves would require minutes, tens of minutes. It often happens that in involved games the players get into time-trouble and have to make fast moves since almost all the time has been used up. Now suppose we let a computer do the investigating of each possible move. A computer would certainly never have time- trouble when it performs thousands of computations every second; all imaginable possibilities would come out almost instantaneously.

That a computer can perform complicated computations you know, but do you know that a computer can play a fairly good game of chess? We will have more to say on that subject a bit later.

The Number of All Possible Chess Games

Let us make an approximate calculation of the total number of different chess games that can ever be played on a chess board. An exact calculation is quite beyond us, but we will acquaint the reader with a very rough estimate of the total number of possible chess games. The Belgian mathematician M. Kraichik makes the following calculation in his book entitled *The Mathematics of Gaines and Mathematical Diversions.*

"In the first move, white has a choice of 20 moves (16 moves for the eight pawns, each of which can move one or two squares and two moves each of the two knights). For each move of white, black can respond with one of the same 20 moves. Combining each move of white with each move of black, we have 20.20 = 400 different games after the first move of each side.

"After the first move, the number of possible new moves increases. For example, if white made the first move P-K4, then it has a choice of 29 moves the next time. The number of possible moves continues to increase. For instance the queen alone, standing on Q5 has a choice of 27 moves (on the assumption that all squares that it can move to are vacant). To simplify matters, let us assume the following average numbers:

20 possible moves for both sides in the first five moves;

30 possible moves for both sides in all subsequent moves.

"Also, we take the average number of moves in a single game to be 40. Then we get the following expression for the number of possible games:

$$(20\ 20)^{5\cdot}\ (30\ 30)^{35}\text{''}.$$

$$(20\cdot 20)^{5}\cdot(30\cdot 30)^{35} = 20^{10}\cdot 30^{70} = 2^{10}\cdot 3^{70}\cdot 10^{80}$$

Replace 2^{10} by the approximation 1000, or 10^3, and express 3^{70} as

$$3^{70} = 3^{68}\cdot 3^2 \approx 10(3^4)^{17} \approx 10\cdot 80^{17} = 10\cdot 8^{17}\cdot 10^{17} = 2^{51}\cdot 10^{18} = 2\cdot(2010)^{5}\cdot 10^{18} \approx 2^{.1015.1018} = 2\ .10^{33}.$$

And so we obtain

$$(20\ 20).\ (30\ 30)^{35} \approx 10^{3}\ 2\ 10^{33}\ 10^{80} = 2\ 10^{116}.$$

This number leaves far behind the legendary number of grains of wheat asked as payment for the invention of chess (that number was a mere $2^{64} - 1 \approx 18 \cdot 10^{18}$). If the whole population of the world played chess round the clock making one move every second, then it would take no less than 10^{100} centuries to exhaust all the games in this marathon of chess!

The Secret of the Chess Machine

You will probably be very surprised to learn that automatic chess-playing machines have been around for quite some time. How can we reconcile this fact with the practically limitless number of combinations of the chess pieces?

Very simply. The machines never did really exist. They were merely a figment of the imagination. A very popular machine was that of the Hungarian mechanician Wolfgang von Kempelen (1734-1804) who demonstrated his machine at the Austrian and Russian courts and then publicly in Paris and London. Napoleon I played a game of chess with it and was confident that he was playing against a machine. In the middle of last century, the famous automatic chess- playing machine came to the United States and finally perished in a fire in Philadelphia.

Other chess-playing machines did not become as famous, but the belief persisted that such machines really could play chess.

In actual fact, not a single chess-playing machine operated automatically. There was always a clever chess master very much alive inside and he did the playing. The famous machine of Kempelen was in the form of an enormous box filled with complex machinery. On top of the box was a chessboard with chess pieces that were moved by the hand of a large doll. Before the game, anyone of the spectators could look inside and convince himself that no one was hidden there. But the trick was that there was still space enough to hide a small-size man (the part was played at one time by such famous chess masters as Johann Allgaier and William Lewis). It may be the hidden player moved from section to section of the huge machine as the public was invited to investigate the mechanisms. Actually, the machinery did not in the least participate in the games and only served to hide the human player.

From all this we can conclude that the number of chess games is for all practical purposes infinite and machines capable of automatically making the most correct moves exist only in the minds of wishful thinkers. There is therefore no need to fear a crisis in the game of chess.

However, recent developments in the computer field cast some doubt on the correctness of our conclusion. There already exist computers that have been programmed to play chess. These are extremely complicated computing machines capable of performing many thousands of operations per second. We have already discussed such computing machines. Now let us see how a computer actually plays chess.

Of course, no computer can do more than carry out operations on numbers. But the computations are carried out by the machine in accordance with a specific scheme of operations, a definite program that has been worked out beforehand.

A chess program is set up by mathematicians on the basis of a definite tactical plan of the game; by tactics we mean a system of rules which for each position permits choosing a unique move, which is the best move from the standpoint of the given tactical plan. Here is an example. To each chess piece is assigned a definite number of points (the value of the piece):

King	+ 200 points
Queen	+ 9 points
Rook	+5 points
Bishop	+3 points
Knight	+3 points
Pawn	+1 point
Lagging Pawn	-0.5 point
Isolated pawn	-0.5 point
Doubled pawn	-0.5 point

Besides that, there are definite values attached to positional advantages (mobility of men, positions closer to the centre than to the edges of the board, and so forth) that are expressed in tenths of a point. Now subtract the total number of points for black from the total for white. The difference obtained gives a certain picture of the material and positional advantage of white over black. If the difference is positive, then white has a better position than black, but if it is negative, then the position is worse.

The computer determines how the difference may change over the next three moves and chooses an optimal variant out of all possible three-move combinations and then prints out "move made".[4] For one move, the computer requires very little time (depending on the type of program and the speed of the computer) so that time-trouble is something that never affects a chess-playing computer.

True, a machine that can "think through" only three moves ahead is a rather weak "player". (Good chess masters think through 10 and more moves ahead in their combinations.) But on the other hand, rapid progress is being made in the computer field and we can expect much better computer players in the near future.

There are a great many more interesting things in computer-played chess, but they are all far beyond the scope of a book like this. In the next chapter we will consider some elementary computer programs.

The Biggest Number Using Only Three Twos

The reader is probably familiar with the biggest number that can be written by means of three digits. You take three nines and arrange them thus:

$$9^{9^{9}}$$

which is the third "superpower" of 9.

This number is so enormous that no comparisons can help to gauge its immensity. The total number of electrons in the visible universe is a pigmy beside this towering giant. In my book Recreational Arithmetic [in Russian there is a discussion of this monster. The reason why I have come back to the problem is that I want to suggest a different one in the same style.

-4- There are other chess tactics too. For example, the program may not provide for all possible retaliatory moves of the opponent but only strong moves, such as check, take, offensive, defensive and the like. Also, relative to some very strong moves of the opponent, the program may provide for more than three moves ahead. A different scale of values for the chess pieces is also possible. The style of play the machine exhibits changes with the tactics chosen.

SOLUTION

Under the fresh impression of the three-storey arrangement of nines, you will probably hurry to put the twos in the same way:

$$2^{2^2}$$

But this time the effect is quite different. The number is very small, even less than 222. Indeed, all we have is 2^4, which is 16.

The truly largest number is not 222 and neither is it 22^2 (or 484) but

$$2^{22} = 41944304$$

This example is very instructive. It shows that in mathematics it is not always wise to proceed by analogy.

Three Threes

PROBLEM

You will probably be more cautious when you approach the following problem: write down the largest number using only three threes and no signs of operations.

SOLUTION

The three-decker arrangement does not produce the desired result either since

3^{3^3} is only 3^{27}, which is less than 3^{33},

which is the largest number and the answer to our problem.

Three Fours

PROBLEM

Write down the largest number using only three fours and dispensing with any operational signs.

SOLUTION

If you try the pattern of the two preceding problems,

$$4^{44}$$

you will be wrong because this time the tower of three fours,

$$4^{4^4}$$

does yield the largest possible number. True enough, $4^4 = 256$, and 4^{256} is greater than 444.

Three identical Digits

Let us take a closer look at this perplexing situation where some figures generate giants in the stacked arrangement, while others produce pigmies. Let us consider the general case.

Use three like figures to write down the largest number without resorting to any operational symbols.

We denote the chosen number by a letter a. Associated with the arrangement

$$2^{22}, 3^{33}, 4^{44}$$

We have the following notation:

$$a^{10a+a} \text{ or } a^{11a}$$

Now the three-tier arrangement looks like this:

$$a^{a^a}$$

Let us now figure out for what value of a the last arrangement depicts a larger number than the first arrangement. Since both expressions are powers with equal integral bases, it follows that the greater quantity corresponds to the greater exponent. Now when is

$$a^a > 11a?$$

Divide both sides of the inequality by a to get

$$a^{a-1} > 11$$

It is easy to see that a^{a-1} is greater than 11 only if a exceeds 3 because

$$4^{4-1} > 11$$

whereas the powers

$$3^2 \text{ and } 2^1$$

are less than 11.

Now it is clear why we were caught unawares when working the earlier problems: one arrangement is best for twos and threes, and an entirely different one for fours and larger numbers.

The Biggest Number Using Four Ones

PROBLEM

Use four ones to write down the largest number (no mathematical symbols are allowed).

SOLUTION

The first thing that comes to mind -1111- is many times smaller than the power

$$11^{11}$$

Hardly anyone will have the patience to multiply out this expression, but we can estimate its value much faster by using tables of logarithms.

The number exceeds 285 thousand million and consequently is some 25 million times greater than the number 1111.

Four Twos to Make the Biggest Number

PROBLEM

Continuing this series of problems, let us now try four twos. What arrangement of four twos depicts the largest number?

SOLUTION

Altogether there are eight possible combinations:

$$2222,\ 222^2,\ 22^{22},\ 2^{222},\ ,\ ,\ ,$$

Which one represents the largest number?

Let us try the top row first, that is, numbers in two stories.

The first one, 2222, is clearly less than any of the others. To compare the next two: 222^2 and 22^{22}, transform the second thus:

$$22^{22} = 22^{2.1} = (22^2)^{11} = 484^{11}.$$

The last number exceeds 222^2 because both the base and the exponent of 484^{11} are greater than in the case of 222^2.

Now compare 22^{22} with the fourth number of the first row, 2^{222}. We replace 22^{22} by a greater number 32^{22} and then show that even this greater number falls short of 2^{222}.

True enough,

$$32^{22} = (2^5)^{22} = 2^{110}$$

which is a smaller number than 2^{222} (the exponent is smaller).

And so we have the largest number in the top row: 2^{222}. That leaves us with five numbers to compare: 2^{222} and the following four numbers:

$$22^{2^2}, 2^{22^2}, 2^{2^{22}}, 2^{2^{2^2}}$$

The last number, equal to 2^{16}, is too small to continue the contest. Now the first number is equal to 22^4 and is less than 32^4 or 2^{20}. It is less than each of the two numbers that come after it. Thus we have three numbers for comparison and each is a power of 2. It is quite evident that the number with the largest exponent is the largest number. Now of the three exponents,

$$222,\ 484 \text{ and } 2^{20+2}\ (= 2^{10.2}.\ 2^2 \approx 10^6.\ 4)$$

the last one is clearly the largest.

And so the largest number that can be depicted by means of four twos is

$$2^{2^{22}}$$

We can get a rough picture of the size of this number without resorting to logarithmic tables if we take the following approximate equality

$$2^{10} \approx 1000.$$

Then

$$2^{22} = 2^{20}.\ 2^2 \approx 4.\ 10^6$$

$$2^{2^{22}} \approx 2^{4000000} > 10^{1200000}$$

Which means the final number has more than a million digits.

Chapter Two

THE LANGUAGE OF ALGEBRA

The Art of Setting up Equations

The language of algebra is equations. Here is what the great Newton wrote in his algebra textbook entitled *Arithmetica Universalis (Universal Arithmetic):* "In order to settle a question referring to numbers or to the abstract relationships between quantities, one needs only to translate the problem from one's own language to the language of algebra." The following is a problem that Newton translated from human terms into the terms of algebra:

In ordinary language:	In the language of algebra:
A merchant has a certain sum of money.	x
During the first year he spent 100 pounds.	$x - 100$
To the remaining sum he then added one third of it.	$(x - 100) + (x - 100)/3 = (4x - 400)/3$
During the next year he again spent 100 pounds.	$(4x - 400)/3 - 100 = (4x - 700)/3$
And increased the remaining sum by one third of it.	$(4x - 700)/3 + (4x - 700)/9 = (16x - 2800)/9$
During the third year he again spent 100 pounds.	$(16x - 2800)/9 - 100 = (16x - 3700)/9$
After he added to the remainder one third of it	$(16x - 3700)/9 + (16x - 3700)/27 = (64x - 14800)/27$
His capital was twice the original amount.	$(64x - 14800)/27 = 2x$

To determine the original capital at the disposal of the merchant, one only has to solve the last equation.

Solving an equation is often a rather simple matter, the real difficulty lies in setting up the equation on the basis of available facts. You have just seen that the art of setting up an equation does indeed reduce to translating from ordinary language to the language of algebra. But the language of algebra is one of few words, and so translating phrases of everyday speech into algebraic terms is often a hard job, as the reader will see from some problems given below where the task is to set up equations of the first degree.

The Life of Diophantus

PROBLEM

Very few facts about the life of the marvelous mathematician of ancient times Diophantus have come down to us. All that we know about him is taken from an inscription made on his tombstone in the form of a mathematical problem. It reads:

In ordinary language:	In the language of algebra:
Traveller! Here rest the ashes of Diophantus. It is a miracle that numbers can measure the length of his life.	x
A sixth portion of it was a beautiful childhood.	$x/6$
After a twelfth part of his life was over, down covered his chin.	$x/12$
A seventh part he spent in childless wedlock.	$x/7$
Five years then passed and he rejoiced in the birth of his first son.	5
whom Fate measured out a joyous and radiant life on this Earth only half of that of his father's	$x/2$
And in deep grief the old man ended his days on Earth, four years after losing his son	$x = (x/6)+ (x/12)+ (x/7)+ 5 + (x/2) + 4$
So how many years did Diophantus live before death overtook him?	

SOLUTION

Solving the equation, we find that x 84 and we also learn that he married at the age of 21, became a father at the age of 38, lost his son when he was 80 and died when he was 84.

The Horse and the Mule

PROBLEM

A horse and a mule, both heavily loaded, were going side by side. The horse complained of its heavy load. 'What are you complaining about?' replied the mule. 'If I take one sack off your back, my load will become twice as heavy as yours. But if you remove one sack from my back, your load will be the same as mine.'

SOLUTION

If I take one sack,	$x - 1$
my load	$y + 1$
will be twice as heavy as yours	$y + 1 = 2(x - 1)$
but if you take one sack from my back,	$y - 1$
your load	$x + 1$
will be the same as mine	$y - 1 = x + 1$

We have reduced the problem to a system of equations in two unknowns:

$$y + 1 = 2(x + 1)$$

$$y - 1 = x + 1$$

or

$$2x - y = 3$$
$$y - x = 2$$

Solving it, we find that $x = 5$ and $y = 7$. The horse was carrying 5 sacks and the mule 7 sacks.

Four Brothers

PROBLEM

Four brothers have 45 rubles. If the money of the first is increased by 2 rubles and the money of the second is decreased by 2 rubles, and the money of the third is doubled, and the money of the fourth is halved, then all of them will have the same amount of money. How much does each have?

SOLUTION

Four brothers have 45 rubles	$x + y + z + t = 45$
If the first brother gets two more rubles,	$x + 2$
The second has two rubles taken away,	$y - 2$
The third has his sum doubled,	$2z$
The fourth has his sum halved,	$t/2$
Then all four brothers will have the same sum of money	$x + 2 = y - 2 = 2z = t/2$

First we split the last equation into three separate equations:

$$x + 2 = y - 2$$

$$x + 2 = 2z$$

$$x + 2 = t/2$$

whence we have

$$y = x + 4,$$

$$z = (x+2)/2$$

$$t = 2x + 4$$

Then, substituting these values into the first equation, we obtain

$$x + x + 4 + (x+2)/2 + 2x + 4 = 45$$

And from this we get $x = 8$. Then we find $y = 12$, $z = 5$, and $t = 20$, which means the four brothers have 8, 12, 5 and 20 rubles.

Two Birds by the Riverside

PROBLEM

Here is a problem of an Arabic mathematician of the 11th century.

There are two palm trees, one opposite the other on each side of a river. One is 30 cubits high, the other 20 cubits. The distance between the foot of each tree comes to 50 cubits. A bird is perched on the top of each tree. All of a sudden, the birds sec a fish come to the surface of the river between the palm trees. They dive at the same time and reach the fish at the same time.

Find the distance between the foot of the taller tree and the fish.

SOLUTION

Using the drawing shown in Fig. 5 and applying the theorem of Pythagoras, we get

$$AB^2 = 30^2 + x^2, AC^2 = 20^2 + (50 - x)^2$$

But AB = AC since both birds covered their distances in the same time. And so

$$30^2 + x^2 = 20^2 + (50 - x)^2$$

Fig. 4

Opening the brackets and simplifying, we obtain a first-degree equation $100x = 2000$, whence $x = 20$. The fish appeared at a distance of 20 cubits from the palm tree which is 30 cubits high.

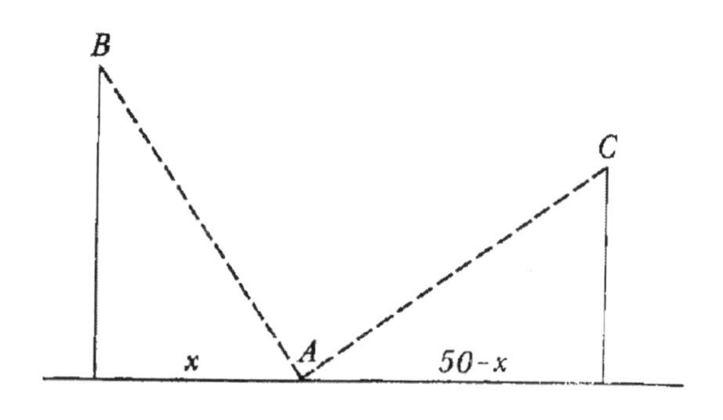

Fig. 5

Out for a Stroll

PROBLEM

"Drop in tomorrow," said the old doctor to his friend.

"Thank you. I'll start out at three. Perhaps you will want to take a walk. If you do, leave at the same time and we'll meet half way."

"You forget that I'm rather old and I can only do 3 km an hour; you are much younger and most likely do 4 km an hour even at a slow pace. You'll have to give me a head start."

"True enough. Since I do one km more per hour than you do, I'll give you a head start of one kilometer, which means I start out a quarter of an hour earlier. Will that be enough?"

"That's very kind of you," the aged doctor replied.

And that was the way it was: the young man started out at two forty-five and walked at 4 km per hour. The doctor left his house at three sharp and did 3 km per hour. When they met, the old doctor turned around and accompanied his young patient to the house.

It was only when the young man got home that he realized that because of the quarter-hour he'd given the doctor, he himself had to cover four times the distance (not twice) that the doctor covered.

What is the distance between the two houses?

SOLUTION

Denote the distance between the two houses by x (km). In all, the young patient walked 2x km and the doctor four times less, or x/2 km. When they met, the doctor had covered half of his distance, or x/4, and the young man the remaining distance, or 3x/4. The doctor walked his distance in x/12 hour, the young man in 3x/16 hour, and we know that he spent 1/4 hour more than the doctor.

This gives us the equation

$$(3x/16) - (x/12) = 1/4$$

Whence x = 2.4 km, zhich is the distance between the young man's house and the doctor's house.

Making Hay

A. V. Tsinger, a prominent physicist, recalls Lev Tolstoy posing the following problem, one that the great writer liked very much.

"A team of haymakers were assigned the task of scything two meadows, one twice the size of the other. Half a day the team worked on the larger meadow. Then it split into two equal groups: the first remained in the larger meadow and finished it by evening; the second group scythed the smaller meadow, but by evening there still remained a portion to do; this portion was scythed the next day by one haymaker in a single day's work.

"How many men were there in the team?"

SOLUTION

Here, besides the chief unknown -the number of men, which we denote by x- it is convenient to introduce another, auxiliary, unknown, namely the area scythed by a single worker in one day; we denote it by y. Although the problem does not require us to find y, it will help us in finding the basic unknown.

Fig. 6

We now express the area of the larger meadow in terms of x and y. This area was worked for half a day x scythemen; they mowed $x.(1/2).\ y = xy/2$.

During the second half of the day, it was worked by half of the team, or x/2 scythemen; they did

$$(x/2).(1/2).y = xy/4.$$

Since the whole meadow was mowed by evening, the total area was

$$xy/2 + xy/4 = 3xy/4.$$

Now let us use x and y to express the area of the smaller meadow. A total of x/2 scythemen worked on it for half a day and mowed an area equal to $(x/2).(1/2).y = xy/4$. Adding the uncut portion, which is equal to y (the area cut by a single scythe man in one working day), we get the area of the smaller meadow:

$$xy/4 + y = (xy + 4y)/4$$

It now remains to put into algebraic language the phrase "the first meadow is twice the size of the second meadow" and we have the equation

$$(3xy/4):(xy + 4y)/4 = 2 \text{ or } 3xy/(xy + 4y) = 2$$

Cancelling y out of the left-hand member of the equation (since y is not needed), we get an equation that looks like this:

$$3x/(x + 4) = 2 \text{ or } 3x = 2x + 8$$

whence $x = 8$.

The team of scythemen consisted of 8 men.

After the first edition of *Algebra Can Be Fun* came out, Professor A. V. Tsinger sent me a detailed and extremely interesting account of the background of this problem. The principal effect of the problem, in his opinion, is that "it is not in the least an algebraic problem, but an arithmetic one and what is more it is very simple, the only difficulty being its unusual form."

Professor Tsinger goes on to describe how the problem originated. "In the period when my father and my uncle 1. 1. Raevsky (a close friend of L. Tolstoy) studied at the mathematics department of Moscow University, there was a subject something like pedagogy. It consisted in students visiting an ordinary city school selected by the University to acquire some teaching experience under the guidance of the best teachers. Now there was a student by the name of Petrov, a friend of Tsinger and liaevsky, and he was an extremely gifted and imaginative fellow. This Petrov (who died at a very early age of tuberculosis, -1 believe) maintained that the children were being spoiled at arithmetic lessons by standard problems and routine methods of solving them. To confirm his belief, Petrov invented problems that were quite out of the ordinary and put the best teachers in a quandary, but were easily solved by capable students who had not yet been spoiled by School. One of these was the problem of the team of scythenien (Petrov thought up a number of such problems). Experienced teachers were able, quite naturally, to solve them with the aid of equations, but a simple arithmetic solution eluded them. Yet, the problem is so simple that there is no need to resort to algebraic methods.

"If the whole team worked on the larger meadow for half a day and half the team for half a day, then it is clear that half the team can cut 1/3 of the meadow in half a day. This means that in the smaller meadow there is an uncut portion equal to 1/2 - 1/3 = 1/6. If one scytheman can work 1/6 of a meadow in one day, and a total of 6/6 + 2/6 = 8/6 was cut, then there must be 8 workers.

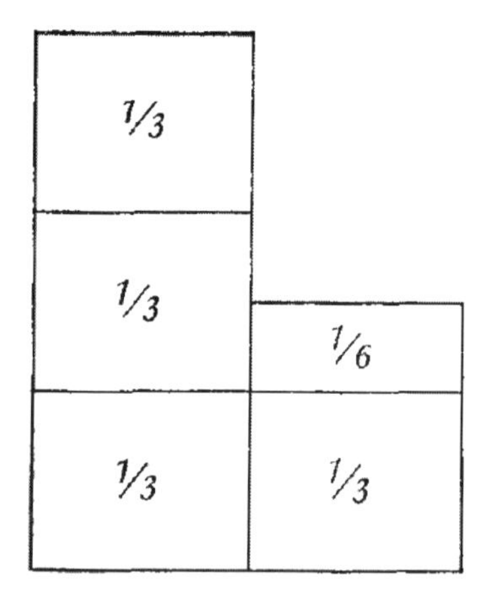

Fig. 7

"Tolstoy, who all his life enjoyed tricky problems that were not too involved, learned about this problem from my father when still a young man. When I met Tolstoy -already an old man- and discussed the problem with him, he was most delighted by the fact that the problem becomes still clearer, literally transparent, if a very simple drawing is employed in the solution (Fig. 7)."

We will now look into several problems that are more easily solved arithmetically than algebraically if one has quick wits.

Cows in the Meadow

PROBLEM

"When studying the sciences, problems are more useful than rules," wrote I. Newton in his *Universal Arithmetic* and accompanied his theoretical propositions with a number of examples. Among these exercises was a problem of pasturing cows, one of a special kind of peculiar problems somewhat like the following.

"The grass in a meadow grew equally thick and fast. It was known that 70 cows could eat it up in 24 days, while 30 cows could do it in 60 days. How many cows would crop the grass of the whole meadow in 96 days?"

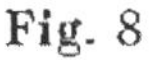

Fig. 8

This problem served as the plot of a humorous story rather reminiscent of Chekhov's *Tutor*. Two grown-ups, relatives of the schoolboy whose task it was to solve the problem, are having a tough time and making little headway.

"This is very strange," says one, "if 70 cows crop the grass in 24 days, then how many will do that in 96 days? Naturally, 1/4 of 70 is 17½ cows... . That's the first piece of nonsense. Here's another piece: 30 cows crop the grass in 60 days; how many cows will do that in 96 days? Still worse, we have 18 ¾ cows. What is more, if 70 cows do the job in 424 days, then 30 cows need 56 days, and not 60, as stated in the problem."

"Have you taken into account that the grass is growing all the time?" asks the other one.

That remark is very much to the point: if the grass is constantly growing and this is disregarded, then it is not only impossible to solve the problem, but even the starting hypothesis will appear to be contradictory.

So how is the problem worked?

SOLUTION

Again we introduce an auxiliary unknown that denotes the daily increase (growth) of grass in fractions of its supply in the meadow. In one day it grows by an amount y. In 24 days it grows 24y. Tithe total amount is, say, 1, then in 24 days the cows will eat up

$$1 + 24y.$$

And in one day, the herd totaling 70 cows will eat

$$(1 + 24y)/24$$

and one cow will eat

$$(1 + 24y)/24.70$$

Likewise, from the fact that 30 cows would have cropped the grass of the meadow in 60 days, we conclude that one cow consumes per day an amount equal to

$$(1 + 60y)/30.60$$

But the amount of grass consumed by a cow in a day is the same for both herds, and so

$$(1 + 24y)/24.70 = (1 + 60y)/30.60$$

whence we get

$$y = 1/480$$

Having found y (the increase), it is easy to determine what portion of the original amount of grass was consumed by one cow in one day:

$$(1 + 24y)/24.70 = [1 + 24\,(1/480)]/24.70 = 1/1600$$

Finally, we set up an equation for the final solution to the problem: if the desired number of cows is x, then

$$[1 + 96.(1/480)]/96x = 1/1600$$

and we have $x = 20$.

Thus, 20 cows would have eaten up all the grass in 96 days.

Newton's Problem

Now let us consider Newton's problem of oxen, after which the preceding one was patterned.

Incidentally, the problem was not devised by Newton himself but is the product of folklore in mathematics.

"Three meadows covered with grass of the same thickness and rate of growth have the following areas: 31/3 hectares, 10 hectares and 24 hectares. The first served to feed 12 oxen during 4 weeks and the second, 21 oxen during 9 weeks. How many oxen can feed on the third meadow in the course of 18 weeks?"

SOLUTION

We introduce an auxiliary unknown *y* to denote what portion of the original supply of grass increases in growth over 1 hectare in one week. In the first meadow, one week sees an increase of 31/3 *y* and in four weeks the grass increases by 3 1/3 *y* = (40/3) *y* of the original supply over one hectare. This is the same as saying that the original area increased to

$$(3\ 1/3 + 40/3\ y)$$

hectares. In other words, the oxen consumed as much grass as covers a meadow with an area of 31/3 + 40/3 y hectares. In one week, 12 oxen consumed a fourth part of this amount, and one ox in one week consumed a 1/48th part, which is the available supply over an area of

$$(3\ 1/3 + 40/3\ y) : 48 = (10 + 40y)/144 \text{ hectares.}$$

In exactly the same way, we find the area that feeds one ox during one week from the data for the second meadow:

One-week growth over 1 hectare = y,

Nine-week growth over 1 hectare = 9y,

Nine-week growth over 10 hectares = 90y.

The area of pasture land containing a supply of grass feeding 21 oxen during 9 weeks is

$$10 + 90y$$

The area sufficient to feed one ox during one week is

$$(10 + 90y)/9.21 = (10 + 90y)/\ 189 \text{ hectares.}$$

Both feeding quotas must be the same:

$$(10 + 40y)/144 = [10 + 40.(1/12)]/114 = 5/54 \text{ hectares.}$$

Finally, we take up the problem proper. Denoting the desired number of oxen by x, we have

$$[24 + 24.18.(1/12)]/18x = 5/54$$

and from this we get x = 36. The third meadow can feed 36 oxen during 18 weeks.

Interchanging the Hands of a Clock

PROBLEM

The biographer and friend of the eminent physicist Albert Einstein, A. Moszkowski, wished to distract his friend during an illness and suggested the following problem (Fig. 9).

The problem he posed was this: "Take the position of the hands of a clock at 12 noon. lithe hour hand and the minute

hand were interchanged in this position the time would still be correct. But at other times (say, at 6 o'clock) the interchange would be absurd, giving a position that never occurs in ordinary clocks: the minute hand cannot be on 6 when the hour hand points to 12. The question that arises is when and how often do the hands of a clock occupy positions in which interchanging the hands yields a new position that is correct for an ordinary clock?

"'Yes,' replied Einstein, 'this is just the type of problem for a person kept to his bed by illness; it is interesting enough and not so very easy. I'm afraid though that the amusement will not last long, because I already have my fingers on a solution'

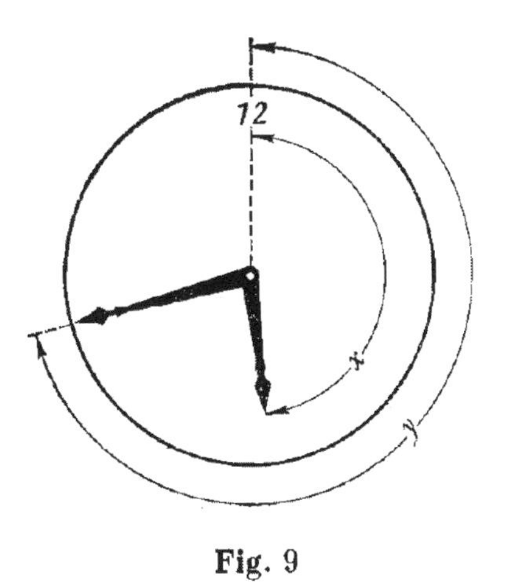

Fig. 9

"Getting up in bed, he took a piece of paper and sketched the hypothesis of the problem. And he solved it in no more time that it took me to state it."

How is this problem tackled?

SOLUTION

Suppose one of the required positions of the hands was observed when the hour hand moved x divisions from 12, and the minute hand moved y divisions. Since the hour hand passes over 60 divisions in 12 hours, or 5 divisions every hour, it covered the x divisions in x/5 hours. In other words, x/5 hours passed after the clock indicated 12 o'clock. The minute hand covered y divisions in y minutes, that is, in y/60 hours. In other words, the minute hand passed the figure 12 a total of y/60 hours ago, or

$$x/5 - y/60$$

hours after both hands stood at twelve. This number is whole (from 0 to 11) since it shows how many whole hours have passed since twelve.

When the hands are interchanged, we similarly find that

$$y/5 - x/60$$

whole hours have passed from 12 o'clock to the time indicated by the hands. This is a whole number (from 0 to 11). And so we have the following system of equations:

$$x/5 - y/60 = m$$

$$y/5 - x/60 = n$$

where m and n are integers (whole numbers) that can vary between 0 and 11. From this system we find

$$x = 60\ (12m + n)/143$$

$$y = 60\ (12n + m)/143$$

By assigning m and n the values from 0 to 11, we can determine all the required positions of the hands. Since each of the 12 values of m can be correlated with each of the 12 values of n, it would appear that the total number of solutions is equal to 12.12 = 144. Actually, however, it is 143 because when in = 0 and n = 0 and also when m = 11 and n = 11 we obtain the same position of the hands.

For m = 11, n = 11 we have

$$x = 60,\ y = 60$$

and the clock reads 1 hour 5 5/11 minutes; the hands have merged by this time and they can of course be interchanged (as in all other cases of coincidence of the hands).

Second example:

$$m = 8,\ n = 5;$$

$$x = 60\ (5 +\ 12.8)/143 \approx 42.38,\ y = 90\ (8 + 12.5)/143 \approx 28.53$$

The respective times are: 8 hours 28.53 minutes and 5 hours 42.38 minutes.

We know the total number of solutions: 143. To find all the points of the dial that yield the required positions of the hands, one has to divide the .circular dial into 143 equal parts, thus obtaining the 143 desired points. At intermediate points no other suitable positions can be found.

The Hands of a Clock Come Together

PROBLEM

How many positions are there on a regular clock with hour hand and minute hand in a coincident position?

SOLUTION

We can take advantage of the equations derived when solving the preceding problem; for if the hour hand and the minute hand can be brought to coincidence, then they can be interchanged, and nothing will change. In this procedure, both hands cover the same number of divisions from the number 12, or $x = y$. Thus, from the reasoning of the preceding problem we can derive the equation

$$x/5 - x/60 = m$$

where m is an integer between 0 and 11. From this equation we find

$$x = 60m/11$$

Of the twelve possible values for m (from 0 to 11) we get 11 (not 12) distinct positions of the hands because when $m = 11$ we find $x = 60$, that is, both hands cover 60 divisions and arrive at 12; the same occurs when in = 0.

Guessing Numbers

The reader is probably familiar with the game of guessing numbers. The conjurer usually suggests performing operations like the following: think up a number, add 2, multiply by 3, subtract 5, subtract the original number and so on -a total of five, even ten operations. He then asks what number you have and, with that answer, he gives you the original number almost at once.

The secret of this "trick" is of course very simple, and again equations give the answer.

Suppose the conjurer suggests a series of operations as indicated in the left-hand column of the following table:

Think up a number	x
add 2	$x + 2$
multiply the result by 3	$3x + 6$
subtract 5	$3x + 1$
subtract the original number	$2x + 1$
multiply by 2	$4x + 2$
subtract 1	$4x + 1$

He then asks you for the final result and gives the answer at once. How does he do this?

It is quite clear from a glance at the right-hand column of the table, where the instructions of the conjurer have been translated into the language of algebra. From this column it is evident that if you think up a number x, then after all the operations you get $4x + 1$. With this knowledge it is easy to "guess" the original number.

Suppose, for example, you got the number 33. What the magician does is solve the equation $4x + 1 = 33$ finding $x = 8$. In other words, he takes the final result, subtracts one ($33-1 = 32$) and then divides that number by 4 to get $32 : 4 = 8$. So the original number was 8. If your final result is 25, the magician does some mental arithmetic ($25 - 1 = 24$, $24 : 4 = 6$) and comes up with the number you thought up, which was 6.

So you see it is very simple. The magician knows beforehand what has to be done with the result in order to obtain the original number.

If that's so, then you can make your friends marvel by letting them suggest the type of operations to be performed on the thought-up number. You suggest that somebody think up a number and perform operations of the following nature in any order: add or subtract a known number (say, add 2, subtract 5, and so on), multiply (but not divide because that will greatly complicate the trick) by a known number (by 2 or 3 and so on), and then add or subtract the original number. To get you completely confused, your friend mounts operation upon operation. Say, he thinks up 5 (which of course is kept secret), and then performs the operations. It goes like this:

"I have thought up a number, multiplied it by 2, added 3, and then added the original number; then I add I, multiply by 2, subtract the original number, subtract 3, and again subtract the original number, and then subtract 2. Finally, I multiply the result by 2 and add 3."

To his great surprise, you fire the answer at him: 5.

How all that is done is now clear enough. When your friend states the , operations he is performing with the original number, you do the same with x. When he says "I have thought up a number" you say (to yourself): "I have x." He says, "I've multiplied it by 2" (and he actually does multiply the original number by 2), whereas you multiply your x by 2 getting 2x. He adds 3 and you follow with 2x + 3, and so forth. When at last he has you in a quandary with his involved operations, you get what is shown in the following table (the left-hand column is what your friend says aloud, and the right-hand column contains the operations that you perform mentally:

I have thought up a number	x
I multiply by 2	2x
add 3 to the result	2x + 3
add the original number	3x +3
add 1	3x + 4
multiply by 2	6x + 8
subtract the original number	5x + 8
subtract 3	5x + 5
again subtract the original number	4x + 5
subtract 2	4x + 3
multiply the result by 2	8x + 6
and add 3.	8x + 9

Then you yourself complete the operations with the result 8x + 9. He says he has 49, which gives you the needed equation: 8 x 9 = 49. To solve it takes a second and you give him the answer straight off: 5.

This is quite a startling trick because you allow your friend to think up any series of operations that he wants to.

True, there is one case when the trick fails. If, for example, after a number of operations you get x + 14, and then your friend says, "and now I subtract the original number and have 14", you follow him with (x + 14) - x = 14 and you do get 14 but there is no equation and you cannot guess the original number. Here is a way out. As soon as you get a result that doe5 not contain the unknown x, you interrupt your friend and say, "Stop! Without any help from you I know what number you now have, it is 14". This is a still greater surprise to your friend because he hasn't told you anything! And although you cannot guess the real (original) number, it is quite a trick after all.

Here is an example of how this is done (your friend's remarks are in the left-hand column as usual):

I have thought up a number	x
I add 2 to it	$x + 2$
multiply the result by 2	$2x + 4$
I now add 3	$2x + 7$
subtract the original number	$x + 7$
Add 5	$x + 12$
And then subtract the original number	12

When you get 12, which is an expression without the unknown x, that is when you have to stop your friend and report the number 12.

All you need is a little practice and you can put on a display of tricks with your friends.

Imaginary Nonsense

PROBLEM

Here is a problem that might appear to be absurd:

What is the number 84 if $8 \cdot 8$ is 54?

This strange question is by no means meaningless, and the problem us solvable with the aid of equations.

Try to figure it out.

SOLUTION

You have probably guessed that the numbers in this problem are not in the decimal system of notation but in some other system, otherwise the phrase "what is the number 84" would indeed be senseless. Suppose the base of the system of numeration is x. Then the number "84" stands for 8 units of the second order and 4 units of the first, or

$$\text{“84”} = 8x + 4.$$

The number "54" means $5x + 4$.

We now have the equation $8.8 = 5x + 4$ or, in the decimal system, $64 = 5x + 4$, whence $x = 12$.

The numbers are written in the duodecimal system of numbers, "84" = $8.12 + 4 = 100$. Thus, if 8.8 = "54", then "84" = 100.

What does 100 stand for if $5.6 = 33$?

Answer: 81 in the nonary (base-9) system of numbers.

The Equation Does the Thinking

If you have ever doubted that an equation can sometimes be cleverer than you yourself, work out the following problem.

The father is 32 years old and his son is 5. How many years will pass before the father is 10 times older than the son? If we denote the sought-for years by x, x years later the father will be $32 + x$ and the son will be $5 + x$. And since the father must be 10 times older than his son, we have the equation

$$32 + x = 10\,(5 + x)$$

which, when solved, yields $x = -2$.

"In minus 2 years" of course simply means two years before. When we set up the equation, we did not give thought to the fact that the age of the father will never be 10 times that of his son in the future; that ratio could only be in the past. The equation this time was a bit wiser than we were and reminded us of our faulty thinking.

Curios and the Unexpected

When working equations we sometimes come up with answers that are a mystery to the inexperienced mathematician. Here are some examples.

I. Find a two-digit number with the following properties. The tens digit is less by 4 than the units digit. If we subtract the desired number from the same digits written in reversed order, we get 27.

Denoting the tens digit by x and the units digit by y, we can readily set up a system of equations for this problem:

$$x = y - 4,$$

$$(10y + x) - (10x + y) = 27.$$

Substituting the value of x from the first equation into the second, we get

$$10y + y - 4 - [10 (y - 4) + y] = 27$$

and, simplifying,

$$36 = 27.$$

We do not yet know the values of the unknowns but we have learned that 36 = 27... . Yet what does that mean?

It merely means that there is no such two-digit number that can satisfy the indicated conditions, and that the equations contradict each other.

Indeed, multiplying both sides of the first equation by 9, we get

$$9y - 9x = 36$$

and from the second equation (after removing brackets and collecting terms) we get

$$9y - 9x = 27.$$

One and the same quantity 9y - 9x is, by tile first equation, equal to 36 and, by the second equation, to 27. This is clearly impossible since 36 71-- 27.

A similar misunderstanding results when solving the following system of equations:

$$x^2y^2 = 8, xy = 4.$$

Dividing the first equation by the second, we get

$$xy = 2$$

and now if we compare the resulting equation with the second one above, we see that

$$xy = 4, \; xy = 2$$

or 4 = 2. There are no numbers that satisfy this system. Such systems of equations that have no solution are said to be inconsistent.

II. A different kind of surprise awaits us if we alter the condition of the preceding problem somewhat. We assume that the tens digit is 3 (not four) less that the units digit, otherwise the problem remains the same. What number is that?

Set up an equation. If the tens digit is denoted by x, then the units digit becomes x + 3. Expressing the rest of the problem in the language of algebra, we get 27 = 27.

This is undoubtedly true, but it tells us nothing about the value of x. Does this mean that there are no numbers that satisfy the requirements of the problem?

Quite the contrary, it means that the equation we set up is an identity: it is true for all values of the unknown x. Indeed, it is easy to see that in this problem every two-digit number in which the units digit exceeds the tens digit by 3 has this property:

$$14 + 27 = 41, \; 47 + 27 - 74,$$

$$25 + 27 = 52, \; 58 + 27 = 85,$$

$$36 + 27 = 63, \; 69 + 27 = 96.$$

III. Find a three-digit number with the following properties:

(1) The tens digit is 7,

(2) The hundreds digit is less by 4 than the units digit,

(3) If the digits of the number are reversed, the new number will exceed the original one by 396.

Let us set up an equation and denote the units digit by x:

$$100x + 70 + x - 4 - [100\,(x - 4) + 70 + x] = 396.$$

After a few simplifications, this equation yields

$$396 = 396.$$

The reader already knows what this means: that every three-digit number in which the first digit is 4 less than the third (the tens digit plays no role) is increased by 396 if the order of the digits is reversed.

Up to now we have been considering problems that are more or less artificial, bookish; their purpose was to help acquire skill in setting up equations. Now that we are theoretically equipped, let us tackle some problems of a practical nature, from industry, everyday life, the military sphere, and sports.

At the Barber's

PROBLEM

Does algebra ever come in handy in cutting hair? Yes it does. I became convinced of this when a barber once approached me with an unusual request:

"Could you please help us here in a problem we have?" "We've spoiled so much of the solution because of it!" put in another barber.

"What is the problem?" I asked.

"We have two solutions of hydrogen peroxide: a 30 per cent solution and a 3 per cent solution. We want to mix them and get a 12 per cent solution, and we can't find the proper proportion."

It turned out to be very simple. What was it?

SOLUTION

It can be solved by arithmetic but algebra does the job faster and more simply. Suppose to make up a 12 per cent mixture we need x grams of the 3 per cent solution and y grams of the 30 per cent solution. Then in the first portion we have 0.03x grams of pure hydrogen peroxide, and in the second 0.3y grams, or altogether

$$0.03x + 0.3y.$$

As a result we have (x + y) grams of the solution in which there must be 0.12 (x + y) grams of pure hydrogen peroxide. We get the equation

$$0.03x + 0.3y = 0.12\,(x + y).$$

From this equation we find x = 2y, which means we have to take twice as much of the 3 per cent solution as of the 30 per cent solution.

Tramcars and a Pedestrian

PROBLEM

While walking along a tram line I noticed that a tramcar caught up with me every 12 minutes, and every 4 minutes a tramcar coming in the opposite direction passed me. We were both (tramcar and I) moving at a uniform rate.

Can you figure out what the time interval is between tramcars leaving their terminals?

SOLUTION

If the tramcars leave their terminals every x minutes, that means that x minutes after I have met a tramcar the following car arrives at that spot. If it is catching up with me, then during the remaining 12 - x minutes it has to cover the same distance that

I do in 12 minutes. Thus, the distance that I cover in one minute is covered by the tram $(12 – x)/12$ minutes.

Now if the tram is coming towards me, then it will meet me 4 minutes after the preceding one, and during the remaining $(x - 4)$ minutes it will cover the same distance that I do in 4 minutes. This means that the distance I cover in one minute is covered by the tram in $(x - 4)/4$ minutes.

We get the equation

$$(12 - x)/12 = (x - 4)/4$$

which yields $x = 6$. The tramcars start out every 6 minutes.

Here is another solution, actually arithmetical. Denote the distance between trams following one another by a. Then the distance between me and the tram coming towards me will diminish by the amount $a/4$ a minute (because together in 4 minutes we cover the distance, equal to a, between the tram that just passed and the next one). Now if a tram is catching up with me, then the distance between us diminishes every minute by $a/12$. Now suppose that I moved ahead during one minute and then turned around and went back one minute (returning to the original spot). During the first minute the distance between me and the tram moving towards me would diminish by $a/4$, and during the second minute (when that same tram was now catching up with me) the distance would diminish by $a/12$. And so in two minutes the distance between us would decrease by $a/4 + a/12 = a/3$. The same would occur if I stood still in one spot, since after all movements I returned to that spot anyway. And so if I didn't move, then in one minute (not in two) the tram would approach me by the amount $a/3 : 2 = a/6$ and the entire distance of a would be covered in 6 minutes. Which means that a tram passes a person standing still every 6 minutes.

Rafts and a Steamboat

PROBLEM

A steamboat covers the distance between a town A and a town B (located downstream) in 5 hours without making any stops. Moving upstream from B to A at the same speed, ircovers the same distance in 7 hours (again making no stops). How many hours does it take a raft moving with the speed of the river current to get from A to B?

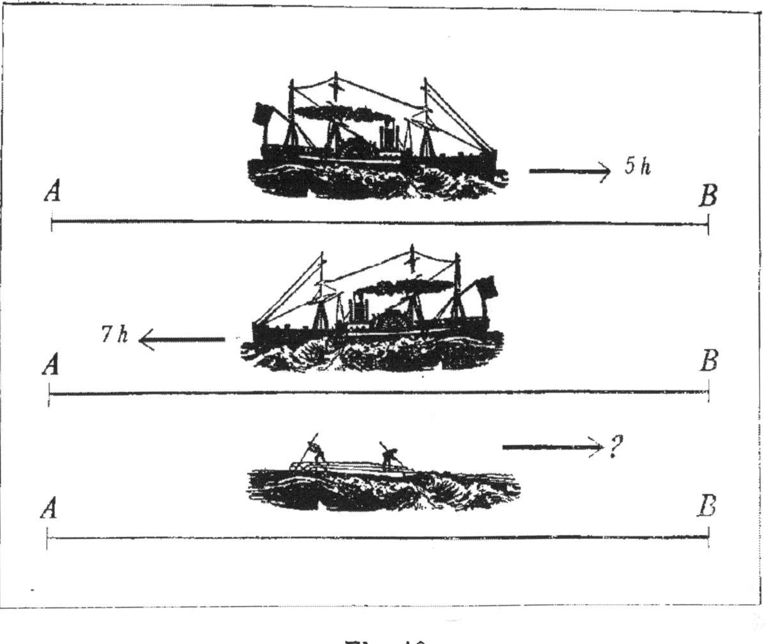

Fig. 10

SOLUTION

We denote by x the time (in hours) it takes the boat to cover the distance between A and B in standing water (at its own speed), and by y the time it takes the rafts to cover that distance. Then the boat does $1/x$ of the AB distance in one hour, and the rafts (going with the current) do $1/y$ of that distance. Therefore when the steamship goes downstream it covers $1/x + 1/y$ of the AB distance, and going upstream (against the current) it does $1/x - 1/y$. Now it is stated in the hypothesis of the problem that going downstream the ship does $1/5$ of the distance in one hour, and going upstream, $1/7$. This gives us the following system of equations:

$$1/x + 1/y = 1/5$$

$$1/x - 1/y = 1/7$$

Note that in solving this system it is best not to get rid of the denominators, simply subtract the second equation from the first. This yields

$$2/y = 2/35$$

and so we get $y = 35$. The rafts cover the distance from A to B in 35 hours.

Two Cans of Coffee

PROBLEM

Two cans containing coffee have the same shape and are made out of the same tin. One can weighs 2 kg and is 12 cm high; the other weighs 1 kg and is 9.5 cm high. Find the net weight of the coffee in both cans.

SOLUTION

Denote the weight of the larger can by x, and that of the smaller one by y. Let the weights of the cans themselves be z and t, respectively. We then have the equations

$$x + z = 2$$

$$y + t = 1$$

Since the weights of the contents of full cans are in the same ratio as their volumes, that is, as the cubes of their heights,[1] it follows that

$$x/y = 12^3/\ 9.5^3 \approx 2.02 \text{ or } x = 2.02\ y$$

Now the weights of the empty cans are in the same ratio as their total surface areas, which is to say, as the squares of their heights. Therefore

$$z/t = 12^2/9.5^2 \approx 1.60 \text{ or } z = 1.60t.$$

Substituting the values of x and z into the first equation, we get the following system:

$$2.02y + 1.60t = 2$$

$$y + t = 1$$

-1- This proportion is applicable only when the tin of the cans is thin. This is because, strictly speaking, the outer and inner surfaces of cans are not similar; what is more, the height of the inside part of a can is, strictly speaking, different from the height of the can itself.

Solving it, we find

$$Y = 20/21 = 0.95, t=0.05$$

And so

$$x = 1.92, z = 0.08.$$

A Question of Dancing

PROBLEM

At a party, 20 people danced. Mary danced with seven partners, Olga with eight, Vera with nine, and so forth up to

Nina who danced with all the partners. Mow many men partners were there at the party?

SOLUTION

This is a very simple problem if the unknown is suitably chosen. Let us seek the number of girls rather than men: the number of girls is x:

1st, Mary danced with 6 +1 partners,

2nd, Olga danced with 6 + 2 partners,

3rd, Vera danced with 6 + 3 partners,

xth, Nina danced with 6 + x partners.

We get the following equation,

$$x+ (6 + x) = 20$$

from which we find that x = 7 and hence that there were 20 - 7 = 13 men at the party.

Reconnaissance at Sea

PROBLEM No. 1

A reconnaissance ship of a squadron is given the assignment to reconnoiter a portion of sea out 70 miles in the direction in which the squadron is moving. The squadron is moving at 35 miles per hour, the reconnaissance ship is doing 70 miles an hour. We have to find out how long it will take the reconnaissance ship to return to the squadron.

SOLUTION

Denote that number of hours by x. During this time the squadron advanced 35x miles and the reconnaissance ship did 70s miles. The reconnaissance ship did 70 miles and then a certain distance on the return leg of its mission, while the squadron simply covered the remaining part of its path. Together, they covered a distance of $70x + 35x$, which is equal to 2.70 miles. This gives us an equation

$$70x + 35x = 140,$$

whence we get

$$x = 140/105 = 1\ 1/3$$

hours. The reconnaissance ship will return to the squadron in 1 hour and 20 minutes.

PROBLEM No. 2

A reconnaissance ship was ordered to reconnoiter a portion of the sea in the direction of motion of the squadron. The ship was to return to the squadron in three hours. How much time will elapse before the reconnaissance ship turns around on the return lap of its mission if it has a speed of 60 knots and the squadron has a speed of 40 knots?

SOLUTION

Suppose the reconnaissance ship has to turn around after x hours; this means it moved x hours away from the squadron, and was on its return mission for a-period of $3 - x$ hours. While all ships were moving in the same direction, the reconnaissance ship, in x hours, covered a distance from the squadron equal to the difference between the distances that they covered, or a total of

$$60x - 40x = 20x.$$

On the return lap, the reconnaissance ship covered a distance, returning to the squadron, of $60 (3 - x)$, whereas the squadron itself did $40 (3 - x)$. Together they did 10s. And so

$$60 (3 - x) + 40 (3 - x) = 20x$$

from which we get

$$x = 2\frac{1}{2}$$

This means that the reconnaissance ship has to turn around 2 hours and 30 minutes after it left the squadron.

At the Cycle Track

PROBLEM

On a circular cycle track there are two cyclists going at unchanging speeds. When they go in opposite directions

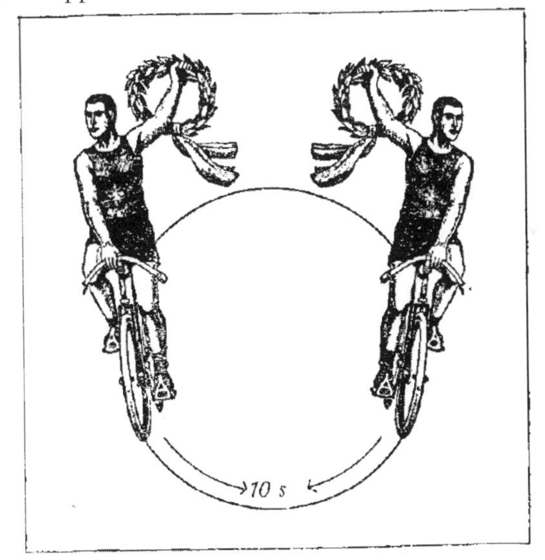

Fig. 11

they meet every 10 seconds; when they go in the same direction, one catches up with the other every 170 seconds. Find the speed of each cyclist if the circular track is 170 meters long.

SOLUTION

If the speed of one cyclist is x, then in 10 seconds he covers a distance of 10x meters. If the other cyclist is moving towards the first one, then during the time between meetings he covers the remaining portion of the circle, or 170 - 10x meters. If the other cyclist has a speed of y, then he does 10y meters in 10 seconds. We have

$$170 - 10x = 10y.$$

Now if the cyclists are following one another, then the first one does 170x meters in 170 seconds, and the other one does 170y meters. If the first one is faster than the second one, he does one complete circle more than the second one before they meet again. We thus have

$$170x - 170y = 170.$$

Simplifying these equations we get

$$x + y = 17, \quad x - y = 1$$

whence

$$x = 9, \quad y = 8 \text{ (meters per second)}$$

A Competition of Motorcyclists

PROBLEM

In a motorcycle competition, one of three motorcycles that started out at the same time was doing 15 km less than the first one and 3 km more than the third, and arrived at the terminal point 12 minutes after the first machine and 3 minutes before the third one. There were no stops en route.

It is required to find:

(1) the length of the course,

(2) the speed of each motorcycle,

(3) the time spent en route by each machine.

SOLUTION

Although we are seeking seven unknowns, we will solve the problem by seeking two: we will set up a system of two equations in two unknowns.

Denote the speed of the second motorcycle by x. Then the speed of the first one is $x + 15$ and of the third $x - 3$. The total path length we denote by y. Then the time en route comes out to

$y/(x + 15)$ for the first motorcycle,

y/x for the second,

$y/(x - 3)$ for the third.

We know that the second motorcycle took 12 minutes (or 1/5 hour) more to cover the distance than the first. And so

$$y/x - y/(x + 15) = 1/5$$

The third motorcycle took 3 minutes (or 1/20 hour) more to cover the whole distance than did the second, and so we have

$$y/(x - 3) - y/x = 1/20$$

Multiply the second of these equations by 4 and subtract from the first to get

$$y/x - y/(x + 15) - 4\,(y/(x-3) - y/x) = 0$$

Divide all terms of this equation by y (this is a quantity that is not zero) and then get rid of the denominators. This yields

$$(x + 15)\,(x - 3) - x\,(x - 3) - 4\,x\,(x + 15) + 4\,(x + 15)\,(x - 3) = 0.$$

Removing brackets and collecting terms, we get

$$3x - 225 = 0$$

and finally

$$x = 75.$$

Knowing x, we can find y from the first equation:

$$y/75 - y/90 = 5$$

or $y = 90$.

So now we have the speeds of the three motorcycles

90, 75, and 72 km per hour.

And the path length is 90 km.

Dividing the path length by the speed of each motorcycle, we can find the time en route:

1 hour for the first motorcycle,

1 h and 12 minutes for the second,

1 h and 15 minutes for the third.

We have thus found all the seven unknowns.

Average Speeds

PROBLEM

An automobile covers the distance between two cities at a speed of 60 km per hour, on the return route the driver does 40 km an hour. Find the average speed.

SOLUTION

The simplicity of the problem is illusive. Without looking deep into the conditions of the problem, many just calculate the average (arithmetic mean) between 60 and 40 and get the half sum, or $(60 + 40)/2 = 50$.

This "simple" solution would be correct if the trip there and back lasted the same time. But it is clear that the return trip (at a smaller speed) must have taken a longer time than the trip there. Taking that into consideration, we can see that 50 is not the answer.

True enough, an equation yields a different answer. It is easy enough to set up an equation if we introduce an auxiliary unknown, namely the quantity l for the distance between the cities. Denoting the sought-for average speed by x, we get the following equation:

$$2l/x = l/60 + l/40$$

Since l is not equal to zero, we can divide through by l and get

$$2/x = 1/16 + 1/40$$

whence

$$x = 48$$

And so the correct answer is 48 and not 50 km an hour. If we worked the problem in literal notation (a for the speed there, and b for the speed on the return trip in kilometers per hour), we would get the equation

$$2l/x = l/a + l/b$$

whence for x we would have

$$2/(1/a + 1/b)$$

This quantity is known as the harmonic mean between a and b.

Thus, the average speed here is not given by the arithmetic mean but by the harmonic mean of the speeds. For positive a and b, the harmonic mean is always less than the arithmetic mean:

$$(a + b)/2$$

As we have just seen in this numerical example (48 is less than 50).

High-Speed Computing Machines

We have been talking about equations and the fun of setting them up and solving problems. Now we bring computers into the discussion. You already know that computers have been taught to play chess (and checkers, or draughts, too). Mathematical machines can also perform such assignments as translating from one language to another (although the result still leaves much to be desired), the orchestration of a piece of music, and much more. The only thing required is to work out a program for the machine to follow.

We will not go into the programs for chess playing or translation from language to language, they are far too complicated, we will analyze only two very simple programs. But first a few words are in order about the construction of a computing machine.

In Chapter One we spoke of machines that are capable of performing many thousands (even millions) of operations a second. The part of the computer that does the actual computations is called the arithmetic unit. Besides that, every computer contains a control unit that organizes the work of the whole machine, and also a memory unit (also called a storage unit). The memory unit is a storage system for storing numbers' and conventional signals. And, finally, the computer is equipped with devices for the input of new digital data and for the output of the final results. These results are in the form of a printout (in the decimal number system) on special cards.

We all know how sound can be recorded on records or on tape and then reproduced. But the recording of sound can be done only once. A new recording requires a fresh disc. A tape recorder performs this task somewhat differently: by magnetizing tape. Here the recorded sound can be reproduced any number of times and if the recording is no longer needed, it can be erased and a new recording made in its place. The same tape may be used for recording many different things, and each time the preceding recording is simply erased.

A similar principle is used in the memory units of computers. Electric, magnetic and mechanical signals are used to record numbers and conventional signals on a special drum, tape or other device, The recorded number can be "read" at any time and if it is no longer needed, it can be erased and replaced by another number. The recording (storing) and reading of numbers or conventional signals takes but millionths of a second.

The memory of a computer may hold several thousand storage locations (memory cells), and each location may have tens of elements (say magnetic elements). In order to write numbers in binary (the binary system of notation), we agree that each magnetized element depicts the digit 1 and each non-magnetized element depicts the digit 0. Suppose each storage location of the memory unit has 25 elements (or, as it is common to say, 25 binary digits); the first digit of the location denotes the sign of the number + or -), the next 14 digits serve to record the integral part of the number, and the last 10 digits record the fractional part of the number. Fig. 12 is a schematic diagram of two storage locations of the memory unit of a computer; each has a storage capacity of 25 digits. The magnetized elements are indicated by ‘H-” signs, the un-magnetized elements by “-” signs. Let us examine the upper storage location (a dot indicates where the fractional part of a number begins and a dashed line separates the first digit, which records the sign of the number, from the other digits). The recorded number reads, in binary, +1011.01 or, in the ordinary decimal system, 11.25.

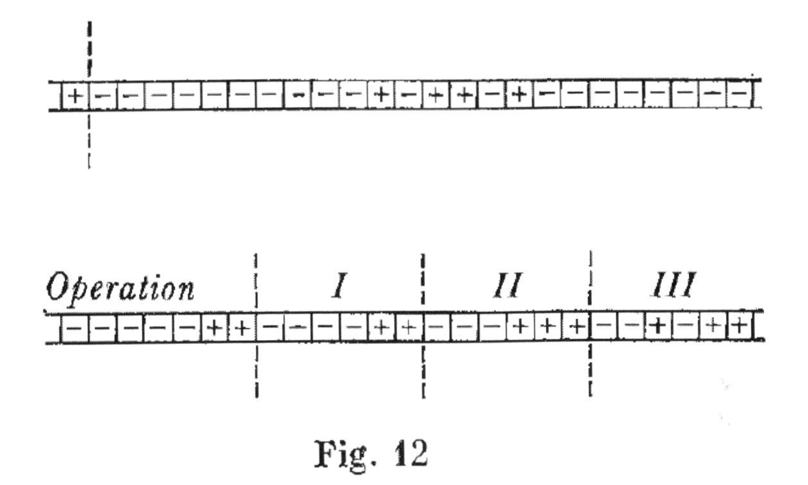

Fig. 12

Memory locations are also used to record instructions *(commands)* that make up the program. Let us see what kind of instructions there are for a so-called *three-address* computer. In this case, to record an instruction the storage location is divided into 4 parts (indicated by dashed vertical lines in the lower storage location in Fig. 12). The first part serves to indicate the operation (operations are recorded in the form of numbers).

For example,

addition - operation 1,

subtraction - operation 2,

Multiplication- operation 3, and so forth.

The instructions are deciphered as follows: the first portion of the location indicates the number of the operation, the second and third parts indicate the numbers of the storage locations (or *addresses)* from which numbers must be extracted in order to perform the operation, and the fourth part indicates the number of the location (the *address)* which is the destination of the final result. For example, in Fig. 12 (bottom row) we have, in binary, the numbers 11, 11, 111, 1011 or, in the decimal system, 3, 3, 7, 11, which means: perform operation 3 (which is multiplication) on the numbers in the *third* and *seventh* memory locations, and then record (store) the result in the *eleventh* location.

From now on we will store the numbers and the instructions in the decimal system directly and not by means of operational symbols as in Fig. 12. For example, the instruction depicted in the lower row of Fig. 12 is written thus:

multiplications 3 7 11

Let us now examine two very simple programs:

Program 1

(1)	addition	4	5	4
(2)	multiplication	4	4	→
(3)	transfer of control	1		
(4)	0			
(5)	1			

Let us see how the computer operates with these data recorded in the first five storage locations in its memory unit.

1st instruction: add the numbers in the 4th and 5th locations and send the result back to the 4th location (in place of what was recorded there earlier). Thus the computer writes the number 0 + 1 = 1 in the 4th location. After the first instruction has been fulfilled, the 4th and 5th locations have the following numbers:

(4) 1,

(5) 1.

2nd instruction: multiply the number in the 4th location by itself (it is squared) and put the result, or 1^2, on a card (the arrow indicates a printout of the final result).

3rd instruction: transfer of control to the 1st location. In other words, the instruction "transfer of control" means that all instructions must again be carried out in order, beginning with the first. So we again have the 1st instruction.

1st instruction: add the numbers in the 4th and 5th locations, and again record the result in the 4th location. Thus in the 4th location we have the number 1 + 1 = 2:

(4) 2,

(5) 1.

2nd instruction: square the number in the 4th location and write out the result, or 2^2, on a card (the arrow indicates a printout).

3rd instruction: transfer of control to the first location (which means a transfer to the 1st instruction again).

(4) 3,

(5) 1.

2nd instruction: print out the number 3^2.

3rd instruction: transfer of control to the 1st location and so on.

We see that the computer computes the *squares of the integers* and prints them out on a card. Note that it is not necessary to write in by hand the next, new number because the machine goes through the sequence of integers and squares each of them. Following that program, the computer finds the squares of all whole numbers, say, from 1 to 10 000 in the course of a few seconds or even fractions of a second.

It must be noted however that actually the program for computing the squares of integers must be somewhat more complicated than that described above. In particular this refers to the second instruction. The point is that printing the result on a card takes much more time than computing a single operation. For this reason, the results are stored in free storage locations of the memory unit and are then later printed out on cards "at leisure", so to say. Thus the first final result is stored in the 1st free location, the second result in the 2nd location, the third in the 3rd location, and so on. This was omitted in the description given above.

Another thing: The computer cannot be engaged for a long time computing squares because there are not enough locations in the memory unit; also there is no way of "guessing" when the machine has computed enough squares to be able to turn it off on time (remember the computer does many thousands of operations per second). To handle this situation, special instructions are included in the program to stop the computer at the right time. For example, the program may state that the computer is to work out the squares of all numbers from 1 to 10 000 and then stop.

There are of course more sophisticated instructions that we will not go into here so as not to complicate our discussion.

Here is a real program for finding the squares of all integers from 1 to 10 000:

Program 1a

(1)	addition		8	9	8
(2)	multiplication		8	8	10
(3)	addition		2	6	2
(4)	conditional jump	8	7	1	
(5)	Stop				
(6)			0	0	1
(7)	10 000				
(8)	0				
(9)	1				
(10)	0				
(11)	0				
(12)	0				

....

The first two instructions don't differ much from those we had in our simplified program. After these two instructions are handled, the 8th, 9th and 10th locations will have the following numbers:

(8) 1

(9) 1

(10) 1^2

The *third instruction* is very interesting: the order is to add the contents of the 2nd and 6th locations and write the results in the 2nd location, which then looks like this:

(2) multiplication 8 8 11.

After the third instruction is carried out, *the second instruction is altered;* to put it more exactly, one of the *addresses* of the 2nd instruction is changed. We will soon learn why that is done.

Fourth instruction: conditional jump (or conditional transfer of control); this is in place of the 3rd instruction in the earlier program. This instruction is performed thus: if the number in the 8th location is *less* than that in the 7th, then control jumps to the 1st locations; otherwise the next (5th) instruction follows. In our case, $1 < 10\,000$ and so the control is transferred to the 1st location. So we again have the 1st instruction.

After the 1st instruction is carried out, 2 appears in the 8th location.

The second instruction will now look like this:

(2) multiplication 8 8 11

and it consists in the number 22 being sent to the 11th location. It is now clear why the 3rd instruction was carried out earlier: the new number, that is 22, does not go to the 10th location, which is already occupied, but to the next one. After instructions one and two have been performed, we get the following numbers:

(8) 2

(9) 1

(10) 1^2

(11) 2^2

After the third instruction, the second location takes on form

(2) multiplication 8 8 12

which means the computer is ready to write the new result in the next, or 12th, location. Since the 8th location still has a smaller number than the 9th location, the 4th instruction is again to transfer control to the 1st location.

Now, after carrying instructions one and two, we get

(8) 3

(9) 1

(10) 1^2

(11) 2^2

(12) 3^2

How long will the computer keep finding the squares of numbers? Until the number 10 000 appears in the 8th location, which means until all the squares of the numbers from 1 to 10 000 have been found. At this point the 4th instruction will not transfer control to the 1st location (this is because the 8th location will have a number that is not less than but equal to the number in the 7th location), and after instruction four is carried out, the computer will take up instruction five: stop (the computer goes off).

Let us now consider a more involved program: the solution of a system of equations. We will examine a simplified version. The reader, if he wishes, can figure out how the complete program would appear.

Given a system of equations:

$$ax + by = c,$$

$$dx + ey = f.$$

This system can readily be solved as

$$x = (ce - bf)/(ae - bd),\ y = (af - cd)/(ae - bd)$$

A few tens of seconds and you will probably be able to solve this system for specified numerical values of the coefficients a, b, c, d, e, f. Now a computer can work out thousands of such systems in one second.

Suppose we have the following program. Given: several systems of equations:

$$ax + by = c,$$

$$dx + ey = f.$$

and

$$a'x + b'y = c',$$

$$d'x + e'y = f'.$$

and

$$a''x + b''y = c'',$$

$$d''x + e''y = f''.$$

with numerical values of the coefficients a, b, c, d, e, f, a', b', c' ,....

Here is an appropriate program:

Program 2

(1) x 28 30 20

(2) x 27 31 21

(3) x 26 30 22

(4) x 27 29 23

(5) x 26 31 24

(6) x 28 29 25

(7) - 20 21 20

(8) - 22 23 21

(9) - 24 25 22

(10): 20 21 →

(11): 22 21 →

(12) + 1 19 1

(13) + 2 19 2

(14) + 3 19 3

(15) + 4 19 4

(16) + 5 19 5

(17) + 6 19 6

(18) Jump 1

(19) 6 6 0

(20) 0

(21) 0

(22) 0

(23) 0

(24) 0

(25) 0

(26) a

(27) b

(28) c

(29) d

(30) e

(31) f

(32) a'

(33) b'

(34) c'

(35) d'

(36) e'

(37) f'

(38) a''

1st instruction: form the product of the numbers in the 28th and 30th locations, then send the result to the 20th location. In other words, in the 20th location we get the number ce.

Instructions two to six are carried out in a similar fashion. After they have been carried out we get the following numbers in locations 20 to 25:

(20) ce

(21) bf

(22) ae

(23) bd

(24) af

(25) cd

Seventh instruction: from the number in the 20th location subtract the number in the 21st location, and again send the result $(cc - bf)$ to the 20th location.

Instructions eight and nine are carried out similarly. Thus, in the locations 20,21, 22 we have the following numbers:

(20) $ce - bf$

(21) $ae - b$

(22) $af - cd$

Instructions 10 and 11: the quotients

$$(ce - bf)/(ae - bd) \text{ and } (af - cd)/(ae - bd)$$

are formed and are printed out on a card (which means they are given as a final result). These are the values of the unknowns obtained from the first system of equations.

The first system has thus been solved. Why are any further instructions needed? The next portion of the program (locations 12 to 19) is needed to prepare the computer for solving the second system of equations. Let us see how this is done. Instructions 10 to 17 consist in the following: to the contents of locations 1 to 6 we add the material in location 19, and the results again remain in locations 1 to 6. Thus, after the 17th instruction has been carried out, the first six locations will look like this:

(1) x 34 36 20

(2) x 33 37 21

(3) x 32 36 22

(4) x 33 35 23

(5) x 32 37 24

(6) x 34 35 25

Instruction 18: transfer of control to location one.

In what way do the new notations in the first six locations differ from the earlier material? In that the first two addresses in these locations have the numbers 32 to 37 instead of 26 to 31. This means that the computer will again perform the same operations, but this time it will take numbers from locations 32 to 37 (instead of from 26 to 31) where the coefficients of the second system of equations are located. The computer thus solves the second system of equations, and then goes on to the third system, and so forth.

From what we have seen it is clear that the most important thing is to write a proper program. By itself the computer is quite helpless and can't do anything. All it can do is carry out a program of instructions. There are programs for computing roots, logarithms, sines, for solving equations of high degrees and many more. There are even programs for playing chess, as we have seen, and for translating from one language into another, albeit rather poorly. A computer can do a lot of things, and of course the more complicated the assignment, the more involved is the program of instructions.

We conclude with a word about so-called compiling routines, which are programs by means of which the computer itself can work out a program for solving some problem. This greatly simplifies the writing of programs, which can often be an extremely time-consuming matter.

Chapter Three

AS AN AID TO ARITHMETIC

It occasionally happens that arithmetic cannot by itself give rigorous proof of certain of its own assertions. In such cases it has to resort to the generalizing techniques of algebra. Propositions of this nature that require the help of algebra include, for example, many of the rules of abridged operations, the curious properties of certain numbers, criteria for divisibility of numbers, and so on. This chapter will be devoted to an examination of a few such problems.

Instantaneous Multiplication

Calculating prodigies (sometimes called lightning calculators) often simplify their computational work by resorting to simple algebraic manipulations. For example, to square 988 one does as follows:

$$988 \,.\, 988 = (988 + 12) \,.\, (988 - 12) + 12^2 = 1000 \,.\, 976 + 144 = 976\,144$$

It is easy to see that the calculator here made use of the familiar algebraic rule

$$a^2 = a^2 - b^2 + b^2 = (a + b)(a - b) + b^2.$$

We can make good use of this rule in oral calculations. For example,

$$27^2 = (27 + 3)(27 - 3) + 3^2 = 729$$

$$63^2 = 66.60 + 3^2 = 3969,$$

$$18^2 = 20.16 + 2^2 = 324$$

$$37^2 = 40.34 + 3^2 = 1369$$

$$48^2 = 50.46 + 2^2 = 2304$$

$$54^2 = 58.50 + 4^2 = 2916$$

To multiply 986 by 997, we do as follows:

$$986 \cdot 997 = (986 - 3) \cdot 1000 + 31 \cdot 4 = 983042.$$

What is this device based on? Write the factors as

$$(1000 - 14) \cdot (1000 - 3)$$

and then multiply the two binomials by the rules of algebra:

$$1000 \cdot 1000 - 1000 \cdot 14 - 1000 \cdot 3 + 14 \cdot 3$$

And now a few more manipulations give us

$$1000\,(1000 - 14) - 1000 \cdot 3 + 14 \cdot 3 = 1000 \cdot 986 - 1000 \cdot 3 + 14 \cdot 3 = 1000\,(986 - 3) + 14 \cdot 3$$

The last line depicts the device that the calculator uses.

Here is a nice way of multiplying two three-digit numbers in which the number of tens is the same, whereas the sum of the units digits comes to 10. For example, to multiply

$$783 \cdot 787$$

do as follows:

$$78 \cdot 79 = 6162,\ 3 \cdot 7 = 21$$

To get

$$616\ 221.$$

The reasoning behind this procedure becomes clear from the following manipulations:

$$(780 + 3)\,(780 + 7) = 780 \cdot 780 + 780 \cdot 3 + 780 \cdot 7 + 3 \cdot 7 = 780 \cdot 780 + 780 \cdot 10 + 3 \cdot 7 = 780(780 + 10) + 3 \cdot 7 = 780 \cdot 790 + 21 - 616\ 200 + 21.$$

Another technique used in such multiplications is still simpler:

$$783.787 = (785 - 2)\,(785 + 2) = 7852 - 4 = 616\ 225 - 4 = 616\ 221.$$

In this example, we had to square the number 785. Here is a fast way to square numbers ending in 5:

35^2; 3.4 = 12. Answer: 1225.

65^2; 6.7 = 42. Answer: 4225.

75^2; 7.8 = 56, Answer: 5625.

The rule here is to multiply the Lens digit by a number that equals that digit plus one; then adjoin 25 to the product.

This device is based on the following reasoning. If the number of tens (the tens digit) is a, then the total number can be depicted thus:

$$10a + 5.$$

The square of this number (the square of a binomial) is

$$100a^2 + 100a + 25 = 100\,a\,(a + 1) + 25.$$

The expression a (a + 1) is the product of the tens digit by the closest greater number. To multiply a number by 100 and add 25 is the same as adjoining 25 to that number.

This same procedure yields a simple method for squaring a number that consists of a whole number and 1/2. For example,

$$(3\ ½)^2 = 3.5^2 = 12{\cdot}25 = 12\ ¼$$

$$(7\ ½)^2 = 56¼\ ,\ (8½)^2 = 72¼ \text{ and so on.}$$

The Digits 1, 5, and 6

You have probably noticed that when multiplying a series of numbers ending in unity or five we get a number ending in that same digit. It is not so well known that the same holds true for the number 6. This means, incidentally, that any power of a number ending in 6 also ends in 6.

For example, $46^2 = 2116$, $46^3 = 97\ 336$.

This peculiarity of 1, 5, and 6 can be explained algebraically. Let us examine 6.

Numbers ending in 6 may be depicted thus:

$$10a + 6,\ 10b + 6 \text{ and so on,}$$

where a and b are whole numbers.

The product of two such numbers yields

$$100ab + 60b + 60a + 36 = 10\cdot(10ab + 6b + 6a) + 30 + 6 = 10\cdot(10ab + 6b + 6a + 3) + 6$$

So we see that the product is made up of a certain number of tens, and the number 6, which, quite naturally, appears at the end.

The same type of proof can be applied to the numbers 1 and 5.

The foregoing permits us to assert that

386^{2567} ends in 6,

815^{23} ends in 5,

491^{1732} ends in 1, and so forth.

The Number 25 and 76

Now there are two-digit numbers that have the same property as 1, 5, and 6. They include the number 25 and -most likely a surprise to most readers- the number 76. Any two numbers ending in 76 yield a product that also ends in 76.

Let us prove this fact. The general expression for such numbers is

$$100a + 76, 100b + 76, \text{ and so on.}$$

We now multiply together two such numbers to get

$10\,000ab + 7\,600b + 7\,600a + 5776 = 10\,000ab + 7\,600b + 7\,600a + 5\,700 + 76 = 100 \cdot (100ab + 76b + 76a + 57) + 76.$

The proposition is established: the product will end in the number 76.

From this it follows that any power of a number ending in 76 will be a number ending in 76:

$$376^2 = 141\,376,\ 576^3 = 191\,102\,976, \text{ and so forth.}$$

Infinite "Numbers"

There are also longer strings of digits that appear at the end of certain numbers and again appear in their product. We will now show that there are an infinity of such strings of numbers.

We know two-digit strings (groups) of digits that have this property: 25 and 76. To find three-digit groups, adjoin in front of 25 or 76 a digit such that the resulting three- digit group of digits has the required property.

What digit should we adjoin to the number 76? Let us denote it by k. Then the desired three-digit number is

$$100k + 76.$$

The general expression for numbers ending in this group of digits is:

$$1000a + 100k + 76,\ 1000b + 100k + 76 \text{ and so on.}$$

Let us multiply together two numbers of this type. We get

$$1\,000\,000ab + 100\,000ak + 100\,000bk + 76\,000a + 76\,000b + 10\,000k^2 + 15\,200k + 5\,776.$$

All terms except the last two end in at least three zeros. For this reason the product ends in 100k + 76 if the difference

$$15\,200k + 5\,776 - (100k + 76) = 15\,100k + 5700 = 15\,000k + 5000 + 100(k + 7)$$

So the sought-for group of digits is of the form 376, which means that any power of 376 will end in 376. An example is

$$376^2 = 141\,376.$$

If we now want to find a four-digit string of digits with the same property, we have to adjoin another digit in front of 376. Denote it by l and we have the following problem: for what l will the product

$$(10\,000a + 1000l + 376)\,(10\,000b + 1000l + 376)$$

end in 10001 + 376? If we remove the brackets in this product and discard all terms ending in four zeros and more, then we get the terms

$$752\,0001 + 141\,376.$$

The product ends in 1000l + 376 if the difference

$$752\,0001 + 141\,376 - (1000l + 376) = 751\,0001 + 141\,000 = (750\,0001 + 140\,000) + 1000\,(l + 1)$$

And so the sought-for four-digit group is 9376.

To this four-digit group we can adjoin another digit by following the reasoning given above. And we get 09 376. Taking another step, we find the group of digits 109 376, then 7 109 376, and so on.

This adjoining of digits on the left can be continued indefinitely. It yields a number with an infinite number of digits:

$$\ldots 7\,109\,376.$$

Such "numbers" can be added and multiplied by the usual rules; this is because they are written from right to left and, as we know, addition and multiplication by columns is also performed from right to left so that in the sum and the product of two such numbers we can compute one digit after another for as many digits as we desire.

$$x^2 = x.$$

Indeed, the square of this "number" (the product of the number into itself) ends in 76 since each factor ends in 76; for the same reason, the square of the written "number" ends in 376; it ends in 9376, and so on. In other words, as we compute one digit after another in the "number" x^2, where x = ... 7 109 376, we will get the same digits as we have in x, so that $x^2 = x$.

We considered groups of digits that end in 76.[1] If similar reasoning is carried out with respect to groups of digits ending in 5, we get the following digit groups:

5, 25, 625, 0625, 90 625, 890 625, 2 890 625 and so on.

This enables us to write down another infinite "number"

... 2 890 625

that likewise satisfies the equation $x^2 = x$. It may be shown that this infinite "number" is equal, as it were, to

$$(((5^2)^2)^2)^2$$

In the language of infinite "numbers" this interesting result can be stated thus: the equation $x^2 = x$ also has (besides the ordinary solutions x= 0 and x = 1) two "infinite" solutions

x = ... 7 109 376 and x = ... 2 890 625

and no other solutions (in the decimal system of notation).[2]

-1- Note that the two-digit group 76 may be found via reasoning similar to that given above: all one needs to do is find the digit to be adjoined on the left to the digit 6 so that the resulting two-digit group has the desired property. Therefore, the "number" ... 7 109 376 can be obtained by adjoining digits to 6 on the left, one after another.
-2- Infinite "numbers" may also be considered in systems of numeration other than the decimal system. Numbers considered in a system of numeration with base p are termed p-adic numbers.

Additional Payment

A PROBLEM OF FOLKLORE

In days of old, two cattle dealers sold a herd of oxen and received as many rubles for each ox as there were oxen in the herd. They then spent this money to buy a herd of sheep at 10 rubles apiece and one lamb. They divided the lot in half and one got an extra sheep and the other took the lamb and received from his companion a certain sum of money. How much did this additional payment come to (it is assumed that it came to a whole number of rubles)?

SOLUTION

This problem is not amenable to immediate translation into the language of algebra, and no equation can be set up for it. So one has to rely on a sort of free-wheeling mathematical reasoning. But algebra gives arithmetic a helping hand here too.

The price of the herd in rubles is a perfect square since the herd was acquired with money obtained from the sale of n oxen at n rubles per ox. One of the two dealers got an extra sheep, which makes the number of sheep odd; also, for this reason, the number of tens in n2 is odd. What is the units digit?

It can be shown that if in a perfect square the tens digit is odd, then the units digit in that number can only be 6.

$$100a^2 + 20ab + b^2 = (10a^2 + 2ab) \cdot 10 + b^2.$$

There are $10a^2 + 2ab$ tens in this number, and then there are some tens in b^2. But $10a^2 + 2ab$ is divisible by 2 and so is an even number. Therefore the number of tens in $(10a + b)^2$ will be odd only if there turns out to be an odd number of tens in b^2. Now recall what b^2 is. This is the square of the units digit, which moans it is one of the following 10 numbers:

$$0, 1, 4, 9, 16, 25, 36, 49, 64, 81.$$

Of these, only 16 and 36 (both ending in 6) have an odd number of *tens,* which means the perfect square

$$100a^2 + 20ab + b^2$$

can have an odd number of tens only if it ends in 6.

It is now easy to find the answer to our problem. It is clear that the lamb went for 6 rubles. Hence the dealer that got the lamb received 4 rubles less than the other one. To make the shares equal, the owner of the lamb is due 2 rubles from his partner.

The additional payment is thus 2 rubles.

Divisibility by 11

Algebra is a great help in finding criteria for the divisibility of a number by some divisor without performing the division. The criteria of divisibility by 2, 3, 4, 5, 6, 8, 9, and 10 are well known. Let us examine divisibility by 11; it is rather simple and of practical value.

Suppose a multidigit number N has a units, b tens, c hundreds, d thousands, and so on, or

$$N = a + 10b + 100c + 1000d + \dots = a + 10(b + 10c + 100d + \dots),$$

where the dots stand for the sum of higher orders. From N we subtract the number $11(b + 10c + 100d + \dots)$, which is a multiple of 11. Then the resulting difference, which is readily seen to be equal to

$$a - b - 10\,(c + 10d + \dots),$$

will have the same remainder upon division by 11 as the number N. Adding to this difference the number $11\,(c+ + 10d + \dots)$, which is divisible by 11, we get the number

$$a - b + c + 10(d + \dots),$$

which also has the same remainder upon division by 11 as N. From it we subtract the number $11(d + \dots)$, which is divisible by 11, and so on. We finally get

$$a - b + c - d + \dots = (a + c + \dots) - (b + d + \dots).$$

From this we get the following criterion for divisibility by11: from the sum of all digits in odd sites, subtract the sum of all digits standing in even sites; if the difference is 0 or a number (positive or negative) divisible by 11, then the number being tested is a multiple of 11; otherwise, our number is not exactly divisible by 11.

Let us test the number 87 635 064:

$$8 + 6 + 5 + 6 = 25,$$

$$7 + 3 + 0 + 4 = 14,$$

$$25 - 14 = 11.$$

Thus, the given number is divisible by 11.

There is another procedure for determining divisibility by 11 which is convenient for numbers that are not very long. It consists in the number under test being split from right to left into groups (or blocks) of two digits each and then the groups being added. If the resulting sum is exactly divisible by 11, then the original number is a multiple of 11, otherwise it is not. Suppose we want to test the number 528. Dividing it into the appropriate groups of two (5/28) and adding the groups, we get

$$5 + 28 = 33.$$

Since 22 is exactly divisible by 11, so also is the number 528:

$$528 : 11 = 48.$$

We now prove this criterion of divisibility. Split a multi-digit number N into groups of digits. We obtain two- digit (or one-digit[3]) numbers which we designate from right to left as a, b, c, and so on, so that the number N can be written as

$$N = a + 100b + 10\,000c + \ldots = a + 100(b + 100c + \ldots).$$

Subtract from N the number 99(b + 100 c + ...), which is divisible by 11. The resulting number

$$a + (b + 100c + \ldots) = a + b + 100\,(c + \ldots)$$

-3- If the number N had an odd number of digits, the last (leftmost) group will be a one-digit block. Besides, a block of the form 03 should also be regarded as a one-digit number 3.

will have the same remainder, when divided by 11, as the number N. From this number we subtract 99 (c + ...) which is divisible by 11, and so on. Finally, we find that the number N has the same remainder upon division by 11 as the number

A License Number

PROBLEM

Three students of mathematics out for a stroll noticed a car break the traffic regulations. Not one of the students noticed the number on the license plate (it was a four-digit number), but, being mathematicians, they noticed some peculiarities about the number. One recalled that the first two digits were the same. A second recalled that the last two digits were also alike. And finally the third student maintained that the four-digit number was a perfect square. Is this information enough to find out the number on the license plate?

SOLUTION

Denote the first (and second) digit of the desired number by a, the third (and fourth) by b. Then we have the number

$$1000a + 100a + 10b + b = 1100a + 11b = 11\,(100a + b).$$

This number is divisible by 11 and so, being a perfect square, is also divisible by 112. Which means the number 100a 1- + b is divisible by 11. Using either one of the two earlier devised criteria for divisibility by 11, we find that 11 divides the number $a + b$. And this means that

$$a + b = 11$$

since each of the digits a, b is less than ten.

The last digit b of the number, which is a perfect square, can assume only the values

$$0, 1, 4, 5, 6, 9.$$

And so for the digit a, which is equal to $11 - b$, we get the following possible values:

$$11, 10, 7, 6, 5, 2.$$

The first two are unsuitable and that leaves us the following possibilities:

$$b = 4, a = 7;$$

$$b = 5, a = 6;$$

$$b = 6, a = 5;$$

$$b = 9, a = 2.$$

We see that the license number can only be:

$$7744, 6655, 5566, 2299$$

But the last three of these numbers are not perfect squares: the number 6655 is divisible by 5 but not by 25; the number 5566 is divisible by 2 but not by 4; the number 2299 = 121 19 is not a square either. That leaves only one number, $7744 = 88^2$, which is the solution.

Divisibility by 19

Justify the following criterion for divisibility by 19. A number is divisible by 19 if and only if the number of tens added to twice the number of units is divisible by 19.

SOLUTION

Any number N may be expressed as

$$N = 10x + y$$

where x is the number of tens (not the tens digit but the total number of integral tens in the whole number), and y is the units digit. We have to show that N is divisible by 19 if and only if

$$N' = x + 2y$$

is a multiple of 19. To do this, multiply N' by 10 and subtract *N* from the product to get

$$10N' - N = 10\,(x + 2y) - (10x + y) = 19y.$$

From this it is clear that if N' is a multiple of 19, then

$$N = 10N' - 19y$$

is exactly divisible by 19; and conversely, if N is exactly divisible by '19, then

$$10N' = N + 19y$$

is a multiple of 19 and then, quite obviously, N' too is exactly divisible by 19.

Apply our divisibility criterion in succession:

$$\begin{array}{r}
4\,704\,588\,|\,1 \\
+2 \\
\hline
47045\,|\,90 \\
+18 \\
\hline
47\,06\,|\,3 \\
+6 \\
\hline
471\,|\,2 \\
+4 \\
\hline
47\,|\,5 \\
+10 \\
\hline
5\,|\,7 \\
+14 \\
\hline
19.
\end{array}$$

Since 19 is exactly divisible by 19, it follows that so also are the numbers 57, 475, 4712, 47 063, 470 459, 4 704 590, 47 045 881.

So the original number is indeed divisible by 19.

A Theorem of Sophie Germain

Here is a problem posed by the eminent French mathematician Sophie Germain.

Prove that every number of the form $a^4 + 4$ is a composite number (provided a is not equal to 1).

SOLUTION

The proof follows from the transformations

$$a^4 + 4 = a^4 + 4a^2 + 4 - 4a^2 = (a^2 + 2)^2 - 4a^2 = (a^2 + 2)^2 - (2a)^2 = (a^2 + 2 - 2a)(a^2 + 2 + 2a).$$

It will be seen that the number $a^4 + 4$ may be expressed as a product of two factors that are not equal to the number itself or to unity[4]; in other words, the number is composite.

Composite Numbers

There are an infinity of the so-called primes (prime numbers), which are integers exceeding unity that cannot be divided by any whole number other than unity and the integer itself.

The sequence of primes begins 2, 3, 5, 7, 11, 13, 17, 19, 23, 29, 31,..., and, as the dots indicate, extends indefinitely. These primes interpose themselves in the range of composite numbers, and split the natural-number sequence into more or less extended portions of composite numbers. How long are these portions? For example, can we find a sequence of, say, a thousand composite numbers with not a single prime among them?

It can be proved, although this may seem improbable, that straight strings of composite numbers between primes may be of *any imaginable length.* There is no bound to the length of such strings: they may consist of a thousand, a million, a trillion and so on composite numbers.

For the sake of convenience, we will make use of the symbol $n!$, which stands for the product of all numbers from 1 to n inclusive. For instance, $5! = 1 \cdot 2 \cdot 3 \cdot 4 \cdot 5$. We will now prove that the sequence

$$[(n + 1)! + 2], [(n + 1)! + 3], [(n + 1)! + 4], \dots \text{ up to } [(n + 1)! + n + 1] \text{ inclusive}$$

consists of n successive composite numbers.

-4- Unity, because $a^2 + 2 - 2a = (a^2 - 2a + 1) + 1 = (a - 1)^2 + 1 \neq 1$, if $a \neq 1$.

These numbers follow one another in the sequence of the natural numbers since each succeeding one is 1 greater than the preceding number. It now remains to prove that they are composite.

The first number

$$(n + 1)! + 2 = 1 \cdot 2 \cdot 3 \cdot 4 \cdot 5 \cdot 6 \cdot 7 \ldots \cdot (n + 1) + 2$$

is even since both terms contain the factor 2. Now, every even number greater than 2 is a composite number. The second number

$$(n + 1)! + 3 = 1 \cdot 2 \cdot 3 \cdot 4 \cdot 5 \ldots (n + 1) + 3$$

consists of two terms, each of which is a multiple of 3. Hence, this number too is composite.

The third number

$$(n + 1)! + 4 = 1 \cdot 2 \cdot 3 \cdot 4 \cdot 5 \ldots (n + 1) + 4$$

is exactly divisible by 4 because it consists of terms that are multiple of 4.

In similar fashion, we find that the number

$$(n + 1)! + 5$$

is a multiple of 5, and so forth. In other words, each number of our sequence contains a factor that is different from unity and from itself; which means it is a composite number.

If you want to write, say, five composite numbers in succession, all you need to do is substitute 5 for n in the sequence given above. You will then get the following sequence:

722, 723, 724, 725, 726.

This is not the only sequence made up of five composite numbers. There are other sequences, such as

62, 63, 64, 65, 66.

Or take the smaller numbers

24, 25, 26, 27, 28.

Let us now try to solve the following problem. Write down a sequence of ten composite numbers.

SOLUTION

Proceed on what has been said. We establish the fact that for the first of the desired ten numbers we can take

$$1 \cdot 2 \cdot 3 \cdot 4 \cdot \ldots \cdot 10 \cdot 11 + 2 = 39\,816\,802.$$

The desired sequence may then look like this:

39 816 802, 39 816 803, 39 816 804 and so on.

There are however sequences of much smaller composite numbers. For instance, there is a sequence of thirteen composite numbers already in the second hundred:

114, 115, 116, 117 and so on up to 126 inclusive.

The Number of Primes

The existence of arbitrarily long sequences of *composite* numbers may seem to suggest that the sequence of *primes* cannot be continued indefinitely. The following proof of the infinitude of prime numbers will clarify this doubt.

The proof belongs to the ancient Greek mathematician Euclid and may be found in his *Elements.* The proof given there is one known as indirect proof, or reductio ad absurdum proof. Suppose the sequence of primes is finite; denote the last prime in the sequence by N. Then form the product

$$1 \cdot 2 \cdot 3 \cdot 4 \cdot 5 \cdot 6 \cdot 7 \ldots N = N!$$

and add unity. This gives us

$$N! + 1.$$

Since this is an integer, it must contain at least one prime factor, which means it must be divisible by at least one prime. But, by hypothesis, the sequence of primes does not exceed N, yet the number N! + 1 cannot be exactly divided by any number less than or equal to N, for every time we obtain a remainder of 1.

Thus, we cannot assume that the sequence of primes is finite: the assumption leads to a contradiction, and this means that no matter how extended the string of composite numbers in the sequence of natural numbers, we may rest assured that at the end of the sequence we will encounter an infinity of primes.

The Largest Prime Discovered So Far

It is one thing to be convinced of the existence of arbitrarily large prime numbers and quite a different thing *to know* exactly what numbers are prime. The larger the natural number, the greater the amount of computation that has to be performed in order to find out whether it is prime or not. The following number is the largest known prime to date:

$$2^{2281} - 1$$

It has about seven hundred digits and a powerful modern computer was used to perform the necessary computations. (See Chapters 1, 2.)

A Responsible Calculation

Mathematical calculations sometimes involve such unwieldy arithmetic that only algebraic methods can save the situation. Suppose we want to find the result of the following operations:

$$2/\ (1+\ 1/90\ 000\ 000\ 000)$$

A word of explanation: this computation is necessary to find out whether engineers who deal with velocities of bodies that are small compared to the speed of propagation of electromagnetic waves can make use of the familiar law of composition of velocities without taking into account the changes brought about by the mechanics of relativity theory. In classical mechanics, a body participating in two motions in the same direction with velocities v_1 and v_2 kilometers per second has a velocity of $(v_1 + v_2)$ kilometers a second. Now the new mechanics gives the velocity of the body as

$$(v_1 + v_2)/\ (1 + v_1 v_2/c^2) \text{ kilometers per second,}$$

where c is the velocity of light in a vacuum (it is equal approximately to 300 000 kilometers a second). To take an example, the velocity of a body taking part in two motions in the same direction, each with a velocity of one kilometer per second, comes out, via the old classical mechanics, to two kilometers per second, and via the new mechanics, to

$$2/(1 + 1/90\,000\,000\,000) \text{ kilometers per second}$$

By how much do these results differ? Is the difference such that it can be detected by extremely sensitive instruments? That is precisely why we have to perform this calculation. We will carry it out in two ways: first in the usual way, arithmetically, and then by means of algebraic procedures. A mere glance at the long rows of figures in the arithmetic approach convinces us of the undoubted advantages of algebra.

To begin with, we manipulate our double-decker fraction into better shape:

$$2/(1 + 1/90\,000\,000\,000) = 180\,000\,000\,000/90\,000\,000\,001.$$

We now carry out the division of the numerator by the denominator:

```
180 000 000 000 | 90 000 000 001
                 ------------------
 90 000 000 001    1.999 999 999 977...
---------------
899 999 999 990
810 000 000 009
 ---------------
 899 999 999 810
 810 000 000 009
  ---------------
  899 999 998 010
  810 000 000 009
   ---------------
   899 999 980 010
   810 000 000 009
    ---------------
    899 999 800 010
    810 000 000 009
     ---------------
     899 998 000 010
     810 000 000 009
      ---------------
      899 980 000 010
      810 000 000 009
       ---------------
       899 800 000 010
       810 000 000 009
        ---------------
        898 000 000 010
        810 000 000 009
         ---------------
         880 000 000 010
         810 000 000 009
          ---------------
          700 000 000 010
          630 000 000 007
          ---------------
           70 000 000 003
```

This is clearly a monotonous, gruelling task where errors can crop up at any stage. Yet it is important when solving this problem to note the exact point at which the sequence of nines is broken and a different sequence of digits sets in.

Now notice how easily algebra handles the situation. It makes use of the following approximate equality: if a is an extremely small fraction, then

$$1/(1 + a) \approx 1 - a$$

where the symbol ≈ stands for "approximately equal to".

It is very easy to see that this assertion holds true: compare the dividend 1 with the product of the divisor by the quotient:

$$1 = (1 + a)(1 - a)$$

or

$$1 = 1 - a^2.$$

Since a is a very small fraction (for example, 0.001), it is clear that a^2 is a still smaller fraction (0.000001) which can be disregarded.

Now let us apply the above to our calculation:[5]

$2/(1+ 1/90\ 000\ 000\ 000) = 2/(1 + 1/9.10^{10}) \approx 2(1 - 0.111\ldots\ 10^{-10}) = 2 - 0.0000000000222\ldots = 1.999999999977\ldots$

The result is the same as we obtained earlier, but the computation is a much shorter one.

The reader is probably curious to learn of what significance this result is in the problem of mechanics posed above.

It shows that due to the smallness of the velocities under consideration as compared with the velocity of light, it is hardly possible to detect any departure from the old law of composition of velocities. Even with such great velocities as one kilometer per second, a difference shows up only in the 11th digit of the number being determined (in ordinary engineering calculations, one confines himself to 4 to 6 digits). We can therefore state very definitely that the new Einsteinian mechanics practically changes nothing in engineering calculations that deal with the "slow" (compared with the velocity of light) bodies. There is, however, a field of modern life where this conclusion calls for caution. It is the field of space flight. Today artificial satellites and

-5- We now make use of the approximate equality $A/(1 + a) \approx A(1 - a)$

space vehicles have reached velocities of the order of 10 km/s. Here the discrepancy between classical and Einsteinian mechanics appears in the ninth digit. And there are higher velocities in the offing…

When It's Easier Without Algebra

Along with cases where algebra is a great aid to arithmetic, there are other cases where it merely complicates matters. A true knowledge of mathematics consists in the ability to deal with mathematical tools so as always to take the straightest and most reliable path, irrespective of whether the procedure is taken from arithmetic, algebra, geometry, or any other branch. It will be useful therefore to examine a case where resorting to algebra can only confuse the solver. The following is an instructive example of such a problem.

Find the smallest of all numbers which when divided

by 2 yield a remainder of 1

by 3 yield a remainder of 2 by 4 yield a remainder of 3

by 5 yield a remainder of 4

by 6 yield a remainder of 5

by 7 yield a remainder of 6

by 8 yield a remainder of 7 by 9 yield a remainder of 8

SOLUTION

I was asked to work out this problem with the words: "How would you go about solving it? There are too many equations; it's easy to get all mixed up."

The trick is simply this: no equations, no algebra -it can be handled very neatly by a simple arithmetical piece of reasoning.

Add one to the desired number. Then what remainder does it yield when divided by 2? The remainder is 1 + 1 = 2; it is exactly divisible by 2.

In the same way, it is exactly divisible by 3, 4, 5, 6, 7, 8, and by 9. The smallest of such numbers is 9.8.7.5 = 2520, and the sought-for number is equal to 2519. This can be tested immediately.

Chapter Four

DIOPHANTINE EQUATIONS

Buying a Sweater

PROBLEM

You have to pay 19 rubles for a sweater. All you have on you are three-ruble bills and the cashier only has five- ruble bills. How can you pay for the sweater? Or maybe you can't?

The question comes down to this: how many three-ruble bills do you have to give to the cashier to pay 19 rubles and receive change from the cashier in the form of five- ruble bills. There are two unknowns in this problem: the number x of three-ruble bills and the number y of five-ruble bills. But there is only one equation that can be set up:

$$3x - 5y = 19.$$

Although one equation in two unknowns has an infinity of solutions, it is not at all obvious that there is even one among them that has integral positive values of x and y (recall that these are the numbers of bills). That is why algebra has worked out a method for solving such indeterminate equations. The credit for introducing them into algebra belongs to the first European representative of that science, the eminent mathematician of antiquity Diophantus, whence the term Diophantine equations.

SOLUTION

We will use the above example to show how such indeterminate equations are solved.

We have to find the values of x and y in the equation $3x - 5y = 19$ knowing that x and y are positive integers (whole numbers).

First we isolate the unknown with the smallest coefficient, the term $3x$, and get $3x= 19 + 5y$

Whence

$$x = (19+ 5y)/3 = 6 + y + (1+2y)/3$$

Since x, 6 and y are integers, the equation can only hold true if $(1+2y)/3$ is a whole number as well. Let us denote it by t. Then

$$x = 6 + y + t$$

where

$$t= (1+2y)/3$$

and, hence,

$$3t = 1 + 2y, 2y = 3t - 1$$

From the latter equation we can determine y:

$$y = (3t-1)/2 = t+ (t-1)/2$$

Since y and t are integers, it follows that (t-1)/2 must likewise be some whole number t_1. Consequently,

$$y = t + t_1$$

and

$$t_1 = (t - 1)/2$$

whence

$$2t_1 = t - 1 \text{ and } t = 2t_1 + 1.$$

Put the value $t = 2t_1 + 1$ into the preceding equations:

$$y = t + t_1 = (2t_1 + 1) + t_1 = 3t_1 + 1,$$

$$x = 6 + y + t = 6 + (3t_1 + 1) + (2t_1 + 1) = 8 + 5t_1$$

And so for x and y we have found the expressions[1]

$$x = 8 + 5t_1,$$

$$y = 1 + 3t_1.$$

Now we know that the numbers *x* and y are not only integers but are also positive, or greater than 0. Consequently,

$$8 + 5t_1 > 0,$$

$$1 + 3t_1 > 0.$$

From these inequalities we find

-1- Strictly speaking, we have only proved that any integer solution of the equation $3x - 5y = 19$ is of the form $x = 8 + 5t_1$, $y = 1 + 3t_1$, where t_1 is some integer. The converse (that is, that for any integer t_1 we obtain some integral solution to the given equation) has not been proved. However, this is easy to see if we reason backwards, so to say, or if we substitute the obtained values of x and y into the original equation.

$$5t_1 > -8 \text{ and } t_1 > -8/5$$

$$3t_1 > -1 \text{ and } t_1 > -1/3$$

Such are the restrictions on t_1: it is greater than -1/3 (and, hence, all the more so greater than -8/5). But since t_1 is a whole number, we conclude that it can be only one of the following values:

$$t_1 = 0, 1, 2, 3, 4, \ldots$$

The corresponding values for x and y are:

$$x = 8 + 5t_1 = 8, 13, 18, 23, \ldots$$

$$y = 1 + 3t_i = 1, 4, 7, 10, \ldots$$

Now at last we have a way to finding out how the payment is to be made:

Either you pay with 8 three-ruble bills and in return receive one five-ruble bill,

$$8.3 - 5 = 19,$$

or you pay 13 three-ruble bills and receive 4 five-ruble bills:

$$13.3 - 4.5 = 19,$$

and so on.

Theoretically, the problem has an infinity of solutions, but in practical situations the number of solutions is limited because neither the buyer nor the cashier has an infinity of bills. For instance, if each has 10 bills, the payment can be made in only one way: by paying 8 three-ruble bills and getting 5 rubles in return. We have thus seen that indeterminate equations are capable of yielding quite definite pairs of solutions in practical problems.

Returning to our problem, we suggest that the reader, as an exercise, work out a variant in which the buyer has only five-ruble bills and the cashier only three-ruble bills. We then get the following series of solutions:

$$x = 5, 8, 11, \ldots$$

$$y = 2, 7, 12, \ldots$$

Indeed,

$$5.5 - 2.3 = 19,$$

$$8.5 - 7.3 = 19,$$

$$11.5 - 12.3 = 19.$$

....

We could obtain these results also from the solution to the main problem by taking advantage of a simple algebraic device. Since giving five-ruble bills and receiving three-ruble bills is the same as receiving negative five-ruble bills and giving negative three-ruble bills, the new version of the problem is solved by means of the same equation that we set up for the main problem:

$$3x - 5y = 19,$$

provided, however, that x and y are negative numbers. Therefore, from the equations

$$x = 8 + 5t_1, \; y = 1 + 3t_1$$

we find (knowing that $x < 0$ and $y < 0$)

$$8 + 5t_1 < 0, \; 1 + 3t_1 < 0$$

and, hence,

$$t_1 < -8/5$$

Assuming $t_1 = -2, -3, -4$ and so forth, we obtain the following values for x and y from the preceding formulas:

$$t_1 = -2, -3, -4,$$

$$x = -2, -7, -1.2,$$

$$y = -5, -8, -11.$$

The first two solutions, $x = -2$, $y = -5$ signify that the buyer pays minus 2 three-ruble bills and receives minus 5 five-ruble bills, or, translated into normal language, he pays 5 five-ruble bills and receives in return 2 three-ruble bills. The same interpretation is applied to the other solutions.

Auditing Accounts

PROBLEM

In auditing the books of a shop, the auditor found that one of the entries was partially blotted out with ink spots and looked like this:

For metres of frieze
at 49r. 36k. a metre r. 28k.

The number of meters sold was blotted out, but obviously it was not in the form of a fraction; the sum of money received was also partially blotted with only the last three digits distinguishable, and it also was evident that there were three preceding figures.

Was it possible for the auditors to recover the original figures from those distinguishable in the entry?

SOLUTION

Let us denote the number of meters by x. The sum of money received for the goods then comes to (in kopecks)

$$4936x.$$

Denote by y the number expressed by the three blotted- out figures in the entry of the total sum of money. This is obviously the number of thousands of kopecks, while the whole sum in kopecks looks like this:

$$1000y + 728.$$

And so we have the equation

$$4936x = 1000y + 728$$

or, dividing through by 8,

$$617x - 125y = 91.$$

In this equation, x and y are whole numbers and y does not exceed 999 since there can only be three figures. Now we solve the equation as indicated above:

$$125y = 617x - 91,$$

$$y = 5x - 1 + (34-8x)/125 = 5x - 1 + 2(17 - 4x)/125 = 5x - 1 + 2t.$$

Here we assumed $617/125 = 5 - 8/125$ because it is best for us to have as small a remainder as possible. The fraction

$$2(17 - 4x)/125$$

is a whole number and since 2 cannot be divided by 125, it follows that $(17-4x)/125$ must be w whole number, which we denote by t.

Then from the equation

$$(17 - 4x)/125 = t$$

we get

$$17 - 4x = 125t,$$

$$x = 4 - 31t + (1-t)/4 = 4 - 31t + t_1$$

where

$$t_1 = (1 - t)/4$$

and, hence,

$$4t_1 = 1 - t,$$

$$t = 1 - 4t_1,$$

$$x = 125t_1 - 27,$$

$$y = 617t_1 - 134.$$ [2]

We know that

$$100 \le y < 1000.$$

-2- Note that the coefficients of t_1 are equal to the coefficients of x and y in the original equation: $617x - 125y = 91$, and the sign is reversed with respect to one of the coefficients of t_1. This is no accident, in fact it may be demonstrated that that is exactly what should occur every time if the coefficients of x and y are relatively prime numbers.

Consequently,

$$100 \leq 617t_1 - 134 < 1000$$

It is clear that t_1 can have only one integral value:

$$t_1 = 1$$

and then

$$x = 98, y = 483,$$

which means that 98 meters were sold for a total of 4837 rubles and 28 kopecks. Thus the entry was restored.

Buying Stamps

PROBLEM

With one ruble it is required to buy 40 stamps in all: one-kopeck, 4-kopeck and 12-kopeck stamps. How many will there be of each denomination?

SOLUTION

Here we have two equations in three unknowns:

$$x + 4y + 12z = 100,$$

$$x + y + z = 40,$$

where x is the number of one-kopeck stamps, y is the number of 4-kopeck stamps, and z the number of 12-kopeck stamps.

Subtracting the second equation from the first, we get one equation in two unknowns:

$$3y + 11z = 60$$

We find y to be

$$y = 20 - 11 \cdot z/3$$

Clearly, $z/3$ is a whole number. Denote it by t. Then we have

$$y = 20 - 11t,$$

$$z = 3t.$$

Substituting the expressions for y and z into the second of the original equations, we get

$$x + 20 - 11t + 3t = 40$$

and

$$x = 20 + 8t.$$

Since $x \geq 0$, $y \geq 0$ and $z \geq 0$, it is easy to see that t has the following bounds:

$$0 \leq t \leq 1\ 9/11$$

and from this we conclude that t can have only two integral values:

$$t = 0 \text{ and } t = 1.$$

The values of x, y and z are then:

t =	0	1
x =	20	28
y =	20	9
z =	0	3

Check

$$20 \cdot 1 + 20 \cdot 4 + 0 \cdot 12 = 100, \quad 28 \cdot 1 + 9 \cdot 4 + 3 \cdot 12 = 100.$$

To summarize, there are only two ways of buying the stamps (and if the requirement is to buy at least one stamp of each denomination, then there is only one way to do that).

Here is another problem of the same vintage.

Buying Fruit

PROBLEM

Five rubles is used to buy 100 items of different kinds of fruit. Here are the prices:

- One water melon 50 kopecks,
- One apple 10 kopecks,
- One plum 1 kopeck.

How many pieces of each type of fruit were bought?

SOLUTION

Denoting the number of water melons by x, the number of apples by y and plums by z, we can set up two equations:

$$50x + 10y + 1z = 500,$$

$$x + y + z = 100.$$

Subtracting the second equation from the first, we get one equation in two unknowns:

$$49x + 9y = 400.$$

The solution continues as

$$y = (400 - 49x)/9 = 44 - 5x + 4(1 - x)/9 = 44 - 5x + 4t$$

$$t = (1 - x)/9,\ x = 1 - 9t$$

$$y = 44 - 5(1 - 9t) + 4t = 39 + 49t.$$

From the inequalities

$$1 - 9t \geq 0 \text{ and } 39 + 49t \geq 0$$

we find that

$$1/9 \geq t \geq -39/49$$

and, hence, $t = 0$. Therefore,

$$x = 1, y = 39.$$

Substituting these values of x and y into the second equation, we get $z = 60$.

No other combinations are possible.

Guessing a Birthday

PROBLEM

The ability to solve indeterminate equations makes it possible to perform the following mathematical trick.

Ask a friend to multiply the number of the date of his birthday by 12, and the number of the month by 31. He reports the sum of both products and you work out the date of his birth.

Suppose your friend was born on February the 9th; then he carries out the following computations:

$$9 \cdot 12 = 108, 2 \cdot 31 = 62,$$

$$108 + 62 = 170.$$

This last number, 170, is what he tells you, and then you work out the date. How?

SOLUTION

The problem reduces to solving the indeterminate equation

$$12x + 31y = 170$$

in positive integers; the day of the month x does not exceed 31, and the number of the month y does not exceed 12.

$$x = (170 - 31y)/12 = 14 - 3y + (2 + 5y)/12 = 14 - 3y + t,$$

$$2 + 5y = 12t,$$

$$Y = (-2 + 12t)/5 = 2t - 2(1 - t)/5 = 2t - 2t_1$$

$$1 - t = 5t_1, t = 1 - 5t_1,$$

$$y = 2(1 - 5t_1) - 2t_1 = 2 - 12t_1,$$

$$x = 14 - 3(2 - 12t_1) + 1 - 5t_1 = 9 + 31t_1.$$

Knowing that $31 \geq x > 0$ and $12 \geq y > 0$, we find the bounds of t_1:

$$-9/31 < t_1 < 1/9$$

Consequently,

$$t_1 = 0, x = 9, y = 2.$$

Another solution that dispenses with equations can also be tried. We are told the number $a = 12x + 31y$. Since $12x + 24y$ is divisible by 12, the numbers $7y$ and a have the same remainders when divided by 12. Multiplying by 7, we find that $49y$ and $7a$ have the same remainders upon division by 12. But $49y = 48y + y$ and $48y$ is divisible by 12. This means y and $7a$ have the same remainders upon division by 12. In other words, if a is not divisible by 12, then y is equal to the remainder upon division of the number $7a$ by 12; but if a is divisible by 12, then $y = 12$. This gives the number of the month y quite definitely. But if we know y then it is easy enough to find x.

A tiny piece of advice: before finding the remainder obtained from the division of $7a$ by 12, replace the number a by its remainder obtained from division by 12. That's much simpler. For example, if $a = 170$, then do the following mental arithmetic:

$$170 = 12 \cdot 14 + 2 \text{ (and so the remainder is 2)},$$

$$2 \cdot 7 = 14;\ 14 = 12 \cdot 1 + 2 \text{ (hence } y = 2\text{)},$$

$$x = (170 - 31y)/2 = (170 - 31 \cdot 2)/12 = 108/12 = 9 \text{ (and so } x = 9\text{)}.$$

You can name the birthday of your friend: it is February 9.

Now let us prove that this trick never fails, in other words, the equation always has only one solution in positive integers. Denote the number given to you by your friend by a so that finding the birthday reduces to solving the equation

$$12x + 31y = a.$$

We assume the contrary. Suppose that the equation has two distinct solutions in positive integers, namely the solution x_1, y_1 and the solution x_2, y_2 (x_1 and x_2 do not exceed 31, and y_1 and y_2 do not exceed 12). We then have

$$12x_1 + 31y_1 = a,$$

$$12x_2 + 31y_2 = a$$

Subtracting the second equation from the first, we get

$$12(x_1 - x_2) + 31(y_1 - y_2) = 0.$$

From this equation it follows that the number $12(x_1 - x_2)$ is divisible by 31. Since x_1 and x_2 are positive numbers that do not exceed 31, their difference $x_1 - x_2$ is less than 31. Therefore the number $12(x_1 - x_2)$ will be divisible by 31 only when $x_1 = x_2$, that is, when the first solution coincides with the second. Thus, the assumption that there are two distinct solutions results in a contradiction.

Selling Chickens

AN OLD PROBLEM

Three sisters came to the market with chickens to sell. One brought 10, another 16 and the third 26. They sold a portion of their chickens by noon at the same price. In the afternoon, afraid that not all the chickens would be sold, they reduced the price and disposed of the remaining stock all at the same price. At the end of the day, each sister had obtained the same amount of money from the sale: 35 rubles. What was the price in the forenoon and in the afternoon?

SOLUTION

Denote the number of chickens sold by each sister in the forenoon by x, y, z. In the afternoon, they sold 10 - x, 16 - y, 26 - z chickens. The earlier price we denote by m, the afternoon price by n. For the sake of clarity, here is a table of the designations:

Number of chickens sold				Price
Forenoon	x	y	z	m
Afternoon	$10 - x$	$16 - y$	$26 - z$	n

The first sister obtained

$$mx + n(10 - x); \text{ hence } mx + n(10 - x) = 35.$$

The second sister obtained

$$my + n(16 - y); \text{ hence } my + n(16 - y) = 35$$

The third sister obtained

$$mz + n(26 - z); \text{ hence } mz + n(26 - z) = 35.$$

A few manipulations yield

$$(m - n)x + 10n = 35$$

$$(m - n)y + 16n = 35$$

$$(m - n)z + 26n = 35$$

Subtracting the first equation from the third and then the second from the third, we get, in succession,

$$(m - n)(z - x) + 16n = 0,$$

$$(m - n)(z - y) + 10n = 0,$$

Or

$$(m - n)(x - z) = 16n,$$

$$(m - n)(y - z) = 10n.$$

Now divide the first of these equations by the second:

$$(x - z)/(y - z) = 8/5 \text{ or } (x - z)/8 = (y - z)/5.$$

Since x, y, z are whole numbers, the differences $x - z$ and $y - z$ are also whole numbers. Therefore, for the equation

$$(x-z)/8 = (y-z)/5$$

to be possible, it is necessary that $x - z$ be divisible by 8, and $y - z$ by 5. Consequently,

$$(x-z)/8 = t = (y-z)/5$$

whence

$$x = z + 8t,$$

$$y = z + 5t.$$

Note that the number t is not only a whole number but is also positive, since $x > z$ (otherwise the first sister would not have been able to net the same sum as the third).

Since $x < 10$, it follows that

$$z + 8t < 10.$$

Given z and t as positive whole numbers, the last inequality is satisfied in only one case: when $z = 1$ and $t = 1$. Substituting these values into the equations

$$x = z + 8t \text{ and } y = z + 5t,$$

we get $x = 9$ and $y = 6$.

Now, going back to the equations

$$mx + n(10 - x) = 35,$$

$$my + n(16 - y) = 35,$$

$$mz + n(26 - z) = 35$$

and substituting into them the values of x, y and z thus found, we obtain the prices at which the chickens were sold:

$$m = 3\tfrac{3}{4} \text{ rubles, } n = 1\tfrac{1}{4} \text{ rubles.}$$

Thus, in the forenoon the chickens sold for 3 rubles and 75 kopecks, and in the afternoon for 1 ruble and 25 kopecks.

Two Numbers and Four Operations

PROBLEM

In solving the preceding problem, which led to three equations in five unknowns, we did not follow a general pattern but rather made use of free mathematical reasoning. We will do the same with respect to the following problems which lead to indeterminate equations of the second degree.

Here's the first.

The following four operations were performed on two positive integers:

(1) they were added,

(2) the lesser was subtracted from the greater,

(3) they were multiplied together,

(4) and the larger number was divided by the smaller one. The results thus obtained were then combined to form a sum of 243, Find the numbers.

SOLUTION

If the greater number is x and the smaller one y, then

$$(x + y) + (x - y) + xy + x/y = 243.$$

If this equation is now multiplied by y, the brackets removed, and like terms collected we get

$$x \cdot (2y + y^2 + 1) = 243y.$$

But $2y + y^2 + 1 = (y+1)^2$ and so

$$x = 243y/(y + 1)^2$$

For x to be a whole number, the denominator $(y + 1)^2$ must be one of the divisors of the number 243 (because y cannot have the same factors as y + 1). Knowing that 24 = 35, we conclude that 243 is divisible only by the following numbers, which are perfect squares: 1, 3^2, 9^2. This means $(y + 1)^2$ must be equal to 1, 3^2 or 9^2, whence, recalling that y must be *positive,* we find that y is equal to 8 or 2.

Then x is equal to

$$243 \cdot 8/81 \text{ or } 243 \cdot 2/9$$

And so the desired numbers are 24 and 8 or 54 and 2.

What Kind of Rectangle?

PROBLEM

The sides of a rectangle are whole numbers. What must their lengths be for the perimeter of the rectangle to be numerically equal to its area?

SOLUTION

Denoting the sides of the rectangle by x and y, we set up the equation

$$2x + 2y = xy$$

Whence

$$x = 2y/(y - 2)$$

Since x and y must be positive, so also must the number y - 2, or y must be greater than 2.

Now notice that

$$x = 2y/(y- 2) = [2(y - 2) + 4]/(y - 2) = 2+ [4/(y - 2)]$$

Since x must be a whole number, the expression $4/(y - 2)$ must be a whole number too. But when $y > 2$, this is only possible if y is equal to 3, 4 or 6. The corresponding values of x are then 6, 4, 3.

To summarize: the sought-for figure is either a rectangle with sides 3 and 6 or a square with side 4.

Two Two-Digit Numbers

PROBLEM

The numbers 46 and 96 are rather peculiar: their product does not change if the digits are interchanged.

Look,

$$46 \cdot 96 = 4416 = 64 \cdot 69.$$

It is required to find out whether there are any other pairs of two-digit numbers having the same property. Is there any way to find them all?

SOLUTION

Denoting the digits of the desired numbers by x and y, z and t, we set up the equation

$$(10x + y)(10z + t) = (10y + x)(10t + z).$$

Removing brackets and simplifying, we get

$$xz = yt,$$

where x, y, z, t are integers less than 10. To find the solutions we set up pairs of equal products made up of 9 digits:

$$1 \cdot 4 = 2 \cdot 2$$

$$1 \cdot 6 = 2 \cdot 3$$

$$1 \cdot 8 = 2 \cdot 4$$

$$1 \cdot 9 = 3 \cdot 3$$

$$2 \cdot 6 = 3 \cdot 4$$

$$2 \cdot 8 = 4 \cdot 4$$

$$2 \cdot 9 = 3 \cdot 6$$

$$3 \cdot 8 = 4 \cdot 6$$

$$4 \cdot 9 = 6 \cdot 6$$

There are nine equalities. From each one it is possible to set up one or two desired groups of numbers. For example, using the equality $1 \cdot 4 = 2 \cdot 2$ we find one solution:

$$12 \cdot 42 = 21 \cdot 24.$$

Using $1 \cdot 6 = 2 \cdot 3$ we get two solutions:

$$12 \cdot 63 = 21 \cdot 36, \; 13 \cdot 62 = 31 \cdot 26.$$

In this manner we obtain the following 14 solutions:

$$12 \cdot 42 = 21 \cdot 24$$

$$12 \cdot 63 = 21 \cdot 36$$

$$12 \cdot 84 = 21 \cdot 48$$

$$13 \cdot 62 = 31 \cdot 26$$

$$13 \cdot 93 = 31 \cdot 39$$

$$14 \cdot 82 = 41 \cdot 28$$

$$23 \cdot 64 = 32 \cdot 46$$

$$23 \cdot 96 = 32 \cdot 69$$

$$24 \cdot 63 = 42 \cdot 36$$

$$24 \cdot 84 = 42 \cdot 48$$

$$26 \cdot 93 = 62 \cdot 39$$

$$34 \cdot 86 = 43 \cdot 68$$

$$36 \cdot 84 = 63 \cdot 48$$

$$46 \cdot 96 = 64 \cdot 69$$

Pythagorean Numbers

A convenient and very exact method used by surveyors to lay down perpendicular lines consists in the following. Suppose it is required to draw a perpendicular to the straight line MN through point A (Fig. 13). To do this, from A lay off a distance a three times on AM. Then make three knots in a rope with distances between knots equal to 4a and 5a. Take the extreme knots and place them on A and B; then take the middle knot and stretch the rope. The result will be a triangle with a right angle at A.

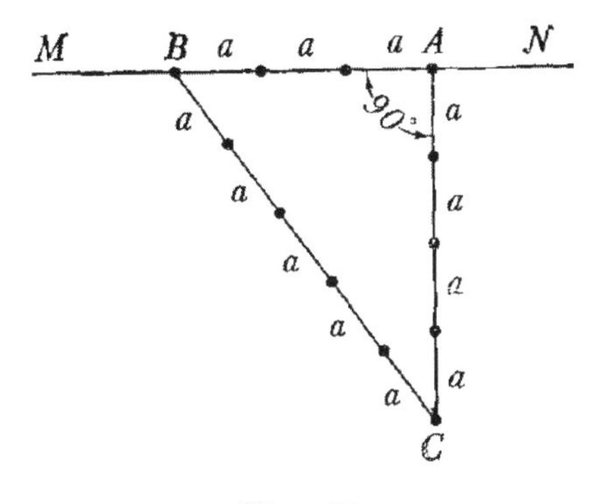

Fig. 13

This ancient procedure, which most likely was used thousands of years ago by the builders of the Egyptian pyramids, is based on the fact that every triangle whose sides are in the ratio 3:4:5 is, by the familiar theorem of Pythagoras, a right-angle triangle because

$$3^2 + 4^2 = 5^2.$$

Besides the numbers 3, 4, 5, there are an infinity of positive integers a, b, c satisfying the relation

$$a^2 + b^2 = c^2$$

They are called *Pythagorean numbers.* According to the *Pythagorean theorem,* such numbers may serve as the sides of a right triangle, and so *a* and *b* are termed the legs and *c* is the hypotenuse.

Clearly, if *a, b, c* is a Pythagorean triad (a triplet of Pythagorean numbers), then *pa, pb, pc,* where *p* is an integral factor, are Pythagorean numbers too. Conversely, if any Pythagorean numbers have a common multiple, then that multiple can be used to divide through all the numbers and again obtain a Pythagorean triad. Therefore, to begin with let us investigate only triplets of relatively prime Pythagorean numbers (the others can be obtained from them by multiplication by an integral factor p).

We will now show that in each of these triplets a, b, c one of the legs must be even and the other odd. We assume the contrary. If both legs a and b are even, then the number $a^2 + b^2$ will be even and hence so also will the hypotenuse. However, this contradicts the fact that the numbers a, b, c do not have any factors in common since three even numbers have the common factor 2. Thus at least one of the legs, a or b, must be odd.

There is still another possibility: both legs are odd and the hypotenuse is even. It is easy to show that this cannot be. Indeed, if the legs are of the form

$$2x + 1 \text{ and } 2y + 1,$$

then the sum of their squares is equal to

$$4x^2 + 4x + 1 + 4y^2 + 4y + 1 = 4(x^2 + x + y^2 + y) + 2,$$

which is a number that, divided by 4, yields 2 as a remainder. Yet the square of any even number should be exactly divisible by 4. This means the sum of the squares of two odd numbers cannot be the square of an even number; in other words, our three numbers are not Pythagorean numbers.

Thus, of the legs a and b, one is even and the other is odd. Therefore, the number $a^2 + b^2$ is odd and, hence, the hypotenuse c is odd.

Suppose, for the sake of definiteness, that leg a is odd and leg b is even. From the equation

$$a^2 + b^2 = c^2$$

we immediately get

$$a^2 = c^2 - b^2 = (c + b)(c - b).$$

The factors $c + b$ and $c - b$ in the right-hand member of the equation are relatively prime. True enough, because if those numbers had a prime factor in common different from unity, then that factor would divide the sum

$$(c + b) + (c - b) = 2c$$

and the difference

$$(c + b) - (c - b) = 2b$$

and the product

$$(c + b)\ (c - b) = a^2,$$

which means the numbers $2c$, $2b$ and a would have a common factor. Since a is odd, the factor is different from 2, and for this reason this same factor is common to the numbers $a, b, c,$ which however cannot be. The contradiction thus obtained shows that the numbers $c + b$ and $c - b$ are relatively prime.

But if the product of relatively prime numbers is a perfect square, then each of them is a square, or

$$c + b = m^2$$

$$c - b = n^2.$$

Solving this system, we get

$$c = (m^2 + n^2)/2,\ b = (m^2 - n^2)/2$$

$$a^2 = (c + b)(c - b) = m^2\ n^2,\ a = mn.$$

And so the Pythagorean numbers under consideration here are of the form

$$a = mn,\ b = (m^2 - n^2)/2,\ c = (m^2 - n^2)/2$$

where m and n are certain relatively prime odd numbers. The reader can readily convince himself of the converse: for any odd m and n the formulas above yield three Pythagorean numbers a, b, c.

Here are several Pythagorean triads obtained for different m and n:

For $m = 3$, $n = 1$, $3^2 + 4^2 = 5^2$

For $m = 5$, $n = 1$, $5^2 + 12^2 = 13^2$

For $m = 7$, $n = 1$, $7^2 + 24^2 = 25^2$

For $m = 9$, $n = 1$, $9 + 40^2 = 41^2$

For $m = 11$, $n = 1$, $11^2 + 60^2 = 61^2$

For $m = 13$, $n = 1$, $13^2 + 84^2 = 85^2$

For $m = 5$, $n = 3$, $15^2 + 8^2 = 17^2$

For $m = 7$, $n = 3$, $21^2 + 20^2 = 29^2$

For $m = 11$, $n = 3$, $33^2 + 56^2 = 65^2$

For $m = 13$, $n = 3$, $39^2 + 80^2 = 89^2$

For $m = 7$, $n = 5$, $35^2 + 12^2 = 37^2$

For $m = 9$, $n = 5$, $45^2 + 28^2 = 53^2$

For $m = 11$, $n = 5$, $55^2 + 48^2 = 73^2$

For $m = 13$, $n = 5$, $65^2 + 72^2 = 97^2$

For $m = 9$, $n = 7$, $63^2 + 16^2 = 65^2$

For $m = 11$, $n = 7$, $77^2 + 36^2 = 85^2$

All other Pythagorean triads either have common factors or contain numbers exceeding a hundred.

Pythagorean numbers have a number of curious properties that we list below without proof:

(1) One of the legs must be a multiple of three.

(2) One of the legs must be a multiple of four.

(3) One of the Pythagorean numbers must be a multiple of five.

A glance at the Pythagorean numbers given above will convince the reader that these properties do hold true.

An Indeterminate Equation of the Third Degree

The sum of the cubes of three integers may be a cube of a fourth number. For instance, $3^3 + 4^3 + 5^3 = 63$.

Incidentally, this means that a cube whose edge is equal to 6 cm is equal in size to the sum of three cubes whose edges are 3 cm, 4 cm and 5 cm (Fig. 14). It is said that this relation highly intrigued Plato.

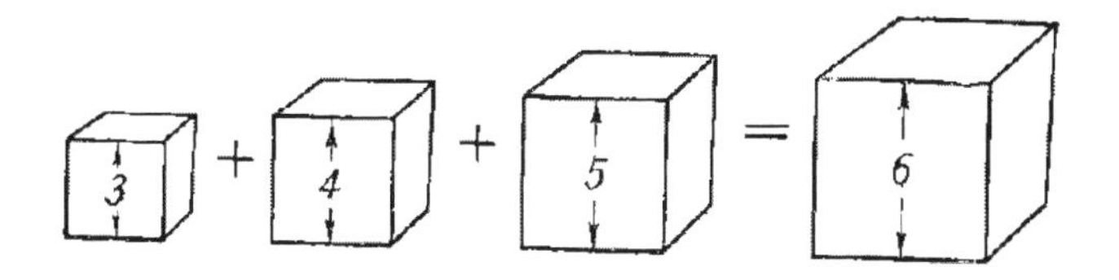

Fig. 14

Let us try to find other relations of the same kind; we pose the problem of finding solutions to the equation

$$x^3 + y^3 + z^3 = u^3$$

It is more convenient to denote the unknown u by -t. Then the equation takes the simple form

$$x^3 + y^3 + z^3 + t^3 = 0.$$

We will now consider a procedure that Will make it possible to find any number (an infinity) of solutions to this equation in integers (positive and negative). Let a, b, c, d and α, β, γ, δ be quadruples of numbers satisfying this equation. To the numbers of the first quadruple add those of the second quadruple multiplied by some number k, and let us also try to choose k so that the resulting numbers

$$a + k\alpha,\ b + k\beta,\ c + k\gamma,\ d + k\delta$$

also satisfy our equation. In other words, we choose k so that the following equality holds true:

$$(a + k\alpha)^3 + (b + k\beta)^3 + (c + k\gamma)^3 + (d + k\delta)^3 = 0.$$

Opening parentheses and recalling that the quadruples a, b, c, d, and α, β, γ, δ satisfy our equation, that is, that we have

$$a^3 + b^3 + c^3 + d^3 = 0, \alpha^3 + \beta^3 + \gamma^3 + \delta^3 = 0,$$

we get

$$3a^2k\alpha + 3ak^2\alpha^2 + 3b^2k\beta + 3bk^2\beta^2 + 3c^2k\gamma + 3ck^2\gamma^2 + 3d^2k\delta + 3dk^2\delta^2 = 0$$

or

$$3k\,[(a^2\alpha + b^2\beta + c^2\gamma + d^2\delta) + k(a\alpha^2 + b\beta^2 + c\gamma^2 + d\delta^2)] = 0.$$

A product vanishes only if at least one of its factors is zero. Equating each of the factors to zero, we obtain two values for k. The first value, $k = 0$, does not interest us: it means that if nothing is added to the numbers a, b, c, d, then the resulting numbers satisfy our equation. So we take only the second value for k:

$$k = (a^2\alpha + b^2\beta + c^2\gamma + d^2\delta)/(a\alpha^2 + b\beta^2 + c\gamma^2 + d\delta^2)]$$

Ii we know two quadruples of numbers that satisfy the original equation, we can find the new quadruple: to do this, add to the numbers of the first quadruple the numbers of the second quadruple multiplied by k, where k has the value given above.

To make use of this procedure, it is necessary to know two quadruples of numbers satisfying the original equation. We already know one: (3, 4, 5, -6). Where can we find another one? This is very simple. For the second quadruple, take the numbers r, -r, s, -s, which obviously satisfy the original equation. In other words, set

$$a = 3, b = 4, c = 5, d = -6,$$

$$\alpha = r, \beta = -r, \gamma = s, \delta = -s.$$

Then, as is easy to see, for k we obtain the following value:

$$k = (-7r - 11s)/(7r^2 - s^2) = (7r + 11s)/(7r^2 - s^2)$$

and the numbers $a + k\alpha$, $b + k\beta$, $c + k\gamma$, $d + k\delta$ will be respectively equal to

$$(28r^2 + 11rs - 3s^2)/(7r^2 - s^2),\ (21r^2 - 11rs - 4s^2)/(7r^2 - s^2),\ (35r^2 + 7rs + 6s^2)/(7r^2 - s^2),\ (-42r^2 - 7rs - 5s^2)/(7r^2 - s^2)$$

By what has already been stated above, these four expressions satisfy the original equation

$$x^3 + y^3 + z^3 + t^3 = 0.$$

Since all these expressions have the same denominator, it can be dropped (which means the numerators of these fractions also satisfy the equation at hand). To summarize, the equation is satisfied (for any r and s) by the following numbers:

$$x = 28r^2 + 11rs - 3s^2,$$

$$y = 21r^2 - 11rs - 4s^2,$$

$$z = 35r2 + 7rs + 6s^2,$$

$$t = -42r^2 - 7rs - 5s^2.$$

One can convince himself directly that this is so by cubing the expressions and adding them. Then, assigning various integral values to r and s, we can obtain a whole series of integral solutions to our equation. If in the process the numbers have a common factor, it can be divided out. For example, for $r = 1$, $s = 1$, we get for x, y, z, t the values 36, 6, 48, -54, or, after dividing through by 6, we have 6, 1, 8, -9. Thus

$$6^3 + 1^3 + 8^3 = 9^3.$$

Here are a number of equalities of his type (obtained after dividing through by a common factor):

For $r = 1$, $s = 2$, $38^3 + 73^3 = 17^3 + 76^3$

For $r = 1$, $s = 3$, $17^3 + 55^3 = 24^3 + 54^3$

For $r = 1$, $s = 5$, $4^3 + 110^3 = 67^3 + 101^3$

For $r = 1$, $s = 4$, $8^3 + 53^3 = 29^3 + 50^3$

For $r = 1$, $s = -1$, $7^3 + 14^3 + 17^3 = 20^3$

For r = 1, s = -2, $2^3 + 16^3 = 9^3 + 15^3$

For r = 2, s = -1, $29^3 + 34^3 + 44^3 = 53^3$

....

Note that if in the original quadruple 3, 4, 5, -6 or in one of the newly obtained quadruples the numbers are transposed and the same procedure is applied, we obtain a new set of solutions. For example, taking the quadruple 3, 5, 4, -6 (that is, by putting a = 3, b = 5, c = 4, d = -6), we get the following values for x, y, z, t:

$$x = 20r^2 + 10rs - 3s^2$$

$$y = 12r^2 - 10rs - 5s^2$$

$$z = 16r^2 + 8rs + 6s^2$$

$$t = -24r^2 - 8rs - 4s^2$$

From this, taking various values of r and s, we obtain the following set of new relations:

for r = 1, s = 1, $9^3 + 10^3 = 1^3 + 12^3$

for r = 1, s = 3, $23^3 + 94^3 = 63^3 + 84^3$

for r = 1, s = 5, $5^3 + 163^3 + 164^3 = 206^3$

for r = 1, s = 6, $7^3 + 54^3 + 57^3 = 70^3$

for r = 2, s = 1, $23^3 + 97^3 + 86^3 = 116^3$

for r = 1, s = -3, $3^3 + 36^3 + 37^3 = 46^3$

and so forth.

In this way we can obtain an infinity of solutions of the equation under consideration.

One Hundred Thousand for the Proof of a Theorem

One of the problems in the field of indeterminate equations became famous through the huge sum of 100 000 German marks that was offered (in a will) for its solution.

The problem is to prove the following proposition that goes by the name of "Fermat's last (or great) theorem".

The sum of identical powers of two integers cannot be the same power of some third integer. The only exception is the second power for which it is possible.

In other words, it is required to prove that the equation

$$x^n + y^n = z^n$$

cannot be solved in whole numbers for $n > 2$.

We have seen that the equations

$$x^2 + y^2 = z^2$$

$$x^3 + y^3 + z^3 = t^3$$

have any number of integral solutions. But try to find three positive integers for which the equation $x^3 + y^3 = z^3$ holds true and all your efforts will be in vain.

It is likewise impossible to find examples for the fourth, fifth, sixth and higher powers. That is the essence of Fermat's great theorem.

Now what is required of seekers of the prize? They have to prove this proposition for all those powers for which it is true. The point is that Fermat's theorem has not yet been proved.

Three centuries have passed since it was first proposed and mathematicians have so far failed to prove it.

The greatest mathematicians have worked on the problem but at best have only proved the theorem for individual exponents or groups of exponents, whereas what is required is a general proof for any integral exponent.

Particularly remarkable is the fact that the proof of the theorem was apparently known at one time, and then was lost. Fermat,[3] the one who proposed the theorem, was an eminent mathematician of the 17th century. He

-3- Fermat (1603 - 1665) was not a professional mathematician. He was educated as a lawyer and was a councillor of the parlianment and his mathematical investigations were done in between. This did not prevent him from making a number of outstanding discoveries, which, incidentally, he did not publish but, as was the custom in those days, described in his letters to scientists and friends: Pascal, Descartes, Huygens, Roberval, and others.

claimed that he knew of a proof. Fermat wrote down his "great" theorem in the margin of a book by Diophantus (as he did a number of other theorems in the field of number theory) and added these words: "I have discovered a truly marvelous demonstration, which this margin is too narrow to contain."

Nowhere has this proof been found, either in the papers of the great mathematician or in his correspondence or anywhere else.

Format's successors were forced to work out the matter in their own way.

Here are the results of these efforts: Euler (1797) proved Format's theorem for the third and fourth powers; the fifth power was proved; by Legendre (1823), the seventh[4] by Lamé and Lebesgue (1840). In 1849 Kummer proved the theorem for a broad range of powers and, incidentally, for all exponents less than one hundred. These latter investigations go far beyond the limits of the realm of mathematics known to Fermat and it is quite a mystery how Fermat could have discovered a general proof of his great theorem. True, he could have been mistaken.

For those interested in the history and the present state of Format's problem we suggest A. Ya. Khinchin's *Fermat's Great Theorem* (in Russian), a nice booklet that can be read by anyone with an elementary knowledge of mathematics.

-4- No special proof is required for composite exponents (except 4): such cases reduce to those of prime exponents.

Chapter Five

THE SIXTH MATHEMATIC OPERATION

The Sixth Operation

Addition and multiplication each have an inverse operation called subtraction and division respectively. The fifth mathematical operation -raising to a power- has two inverses: finding the base and finding the exponent. Finding the base is the sixth mathematical operation and is called extraction of roots. Finding the exponent (this is the seventh operation) is termed taking logarithms. The reason why raising to a power has two inverses, whereas addition and multiplication have only one each is easy to see: both terms in addition (first and second) are of an equal status and can be interchanged. The same goes for multiplication. Now the numbers that take part in raising to a power are not of the same status and, generally, cannot be interchanged (for example $3^5 \neq 5^3$). For this reason, finding each of the numbers participating in addition and multiplication is handled by the same procedures, whereas finding the base of a power and finding the exponent are handled in different ways.

The sixth operation (root extraction) is denoted by the symbol √. It is not so commonly known that this is a modification of the Latin letter r, the initial letter in the word radix, meaning root. There was a time (in the 16th century) when the root symbol was designated by a capital R alongside which stood the first letter of the Latin words quadratus (q) and cubus (c) to -indicate which particular root was being sought.[1] For example, one wrote

R.q. 4352

to mean

$$\sqrt{4352}.$$

Add to this the fact that at that time there were no signs to indicate plus and minus (the letters p. and m. were used instead) and also the fact that our brackets (parentheses) were indicated by the symbols ⌞ and ⌟ and it will be clear that those notations were a far cry from customary algebraic symbols in use today.

Here is an example taken from a book by the mathematician Bombelli (1572):

-1- Magnitsky's textbook of mathematics, which was widely used in the first half of the 18th century in Russia, does not have a special symbol for root extraction.

$$R.c. \llcorner_{R.q.} 4352\ p.\ 16\lrcorner\ _{m.R.c.} \llcorner_{R.q.} 4352\ m.\ 16\lrcorner$$

Using modern symbols, it would look like this:

$$\sqrt[3]{\sqrt{4352}+16}-\sqrt[3]{\sqrt{4352}-16}.$$

Besides the notation we can also make use of , which is particularly convenient in the sense of generalization, for it demonstrates very pictorially that every root is nothing other than a power whose exponent is a fraction. It was proposed by the well-known Flemish engineer and mathematician of the 16th century Simon Stevin.

Which Is Greater?

PROBLEM

Which is greater $\sqrt[5]{5}$ or $\sqrt{2}$?

This and the problems that follow are to be solved *without bothering to compute the values of the roots.*

SOLUTION

Raising both expressions to the 10th power, we get

$$(\sqrt[5]{5})^{10}=5^2=25, \quad (\sqrt{2})^{10}=2^5=32$$

And since 32 > 25, it follows that

$$\sqrt{2}>\sqrt[5]{5}$$

PROBLEM 2

Which is greater $\sqrt[4]{4}$ or $\sqrt[7]{7}$?

SOLUTION

Raising both expressions to the 28th power, we get

$$(\sqrt[4]{4})^{28} = 4^7 = 2^{14} = 2^7 \cdot 2^7 = 128^2,$$
$$(\sqrt[7]{7})^{28} = 7^4 = 7^2 \cdot 7^2 = 49^2.$$

Since 128 > 49, it follows that

$$\sqrt[4]{4} > \sqrt[7]{7}.$$

PROBLEM 3

Find the greater expression of $\sqrt{7} + \sqrt{10}$ and $\sqrt{3} + \sqrt{19}$.

SOLUTION

Squaring both expressions, we get

$$(\sqrt{7} + \sqrt{10})^2 = 17 + 2\sqrt{70}$$

$$(\sqrt{3} + \sqrt{19})^2 = 22 + 2\sqrt{57}$$

Reduce both expressions by 17 and we have

$$2\sqrt{70} \text{ and } 5 + 2\sqrt{57}$$

Square these expressions. This yields

$$280 \text{ and } 253 + 20\sqrt{57}.$$

Subtract 253 from each and then compare them:

$$27 \text{ and } 20\sqrt{57}.$$

Since $\sqrt{57}$ is greater than 2, it follows that $20\sqrt{57} > 40$; hence

$$\sqrt{3} + \sqrt{19} > \sqrt{7} + \sqrt{10}$$

Solve It at a Glance

PROBLEM

Take a close look at the equation

$$x^{x^3} = 3$$

and find x immediately.

SOLUTION

Anyone familiar with algebraic symbols will figure out that

$$x = \sqrt[3]{3}.$$

It must be that because

$$x^3 = (\sqrt[3]{3})^3 = 3$$

and consequently

$$x^{x^3} = x^3 = 3,$$

which is what we sought.

For those who cannot do the problem "at a glance", try this device.

Let

$$x^3 = y.$$

Then

$$x = \sqrt[3]{y}.$$

and the equation becomes

$$(\sqrt[3]{y})^y = 3$$

or, cubing,

$$y^y = 3^3$$

It is clear that y = 3 and, hence,

$$x = \sqrt[3]{y} = \sqrt[3]{3}.$$

Algebraic Comedies

PROBLEM

The sixth mathematical operation makes it possible to devise actual algebraic comedies and farces on such topics as 2·2 = 5, 2 = 3 and the like. The humor in these mathematical shows lies in the fact that the error -a rather elementary one- is somewhat camouflaged and is not at once apparent. Let us take two plays from this comic repertoire from the field of algebra.

To start with,

$$2 = 3.$$

We begin with the un-questionable equality

$$4 - 10 = 9 - 15.$$

Then to both sides of the equality we add the same quantity, 6¼:

$$4 - 10 + 6\tfrac{1}{4} = 9 - 15 + 6\tfrac{1}{4}$$

The comedy goes on with the following manipulations:

$$2^2 - 2 \cdot 2 \cdot 5/2 + (5/2)^2 = 3^2 - 2 \cdot 3 \cdot 5/2 + (5/2)^2,$$

$$(2 - 5/2)^2 = (3 - 5/2)^2$$

Taking the square root of both members of the equation, we get

$$2 - 5/2 = 3 - 5/2$$

Finally, adding 5/2 to both sides, we arrive at our comical result:

$$2 = 3.$$

Where is the mistake?

SOLUTION

An error slipped in when we concluded that from

$$(2 - 5/2)^2 = (3 - 5/2)^2$$

Follows

$$(2 - 5/2) = (3 - 5/2)$$

From the fact that the squares are equal it does not at all follow that the first powers are equal. Say, $(-5)^2 = 5^2$, but -5 does not equal 5. Squares may be equal even when the first powers have different signs. That precisely is the case in our problem:

$$(-\tfrac{1}{2})^2 = (\tfrac{1}{2})^2$$

But -½ is not the same as ½.

PROBLEM 2

Here's another algebraic farce (Fig. 15):

$$2 \cdot 2 = 5.$$

It follows the path of the preceding problem and is based on the same trick. We start out with the undoubtedly flawless equality

$$16 - 36 = 25 - 45.$$

Equal numbers are added to each side:

$$16 - 36 + 20\tfrac{1}{4} = 25 - 45 + 20\tfrac{1}{4}$$

and then the following manipulations are carried out:

$$4^2 - 2\cdot 4\cdot(9/2) + (9/2)^2 = 5^2 - 2\cdot 5\cdot(9/2) + (9/2)^2$$

$$(4 - 9/2)^2 = (5 - 9/2)^2$$

Again using the false conclusion of the earlier problem, we finally get

$$4 - 9/2 = 5 - 9/2,$$

$$4 = 5$$

$$2.2 = 5$$

These amusing instances should be a warning to the inexperienced mathematician in performing imprudent operations with equations involving the radical sign.

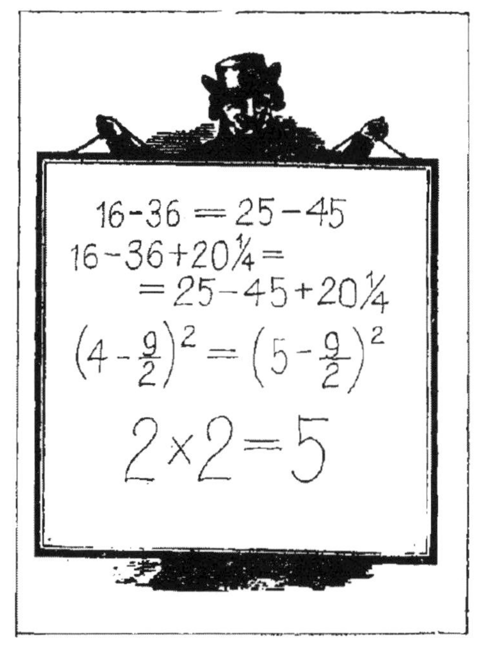

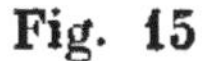
Fig. 15

Chapter Six

SECOND-DEGREE EQUATIONS

Shaking Hands

PROBLEM

A meeting gathered and someone counted the total number of handshakes to be 66. How many people were there at the meeting?

SOLUTION

Algebraically, the problem is solved with great ease. Each of x people shook the hand of x - 1 persons, which puts the total number of handshakes at x (x - 1). But also bear in mind that when Ivanov shakos the hand of Petrov, Petrov also shakes Ivanov's hand. We count these two handshakes as one. That makes the number of handshakes half of x (x - 1):

$$x (x-1)/2 = 66,$$

or, after some simplifying manipulations,

$$x^2 - x - 132 = 0,$$

whence

$$x = (1 \pm \sqrt{(1+528)})/2$$

$$x_1 = 12, x_2 = -11.$$

Since the negative solution (-11 persons) is meaningless here, we discard it and retain only the first root. Twelve persons were at the meeting.

Swarms of Bees

PROBLEM

In ancient India there was a curious kind of sporting contest, a sort of a public competition in the solution of difficult problems. Hindu mathematics manuals served in part as an aid to competitors of such contests in mental sport. One author of such a manual wrote: "The rules given here can be used by a wise man to think up thousands of other problems. Just as the sun in its brightness eclipses the stars, so a wise man can eclipse the glory of another in congregations of people by submitting and solving algebraic problems." The original is more poetic since the whole book is made up of versos, and the problems too arc in verse form. Here is one translated into prose.

A group of bees equal in number to the square root of half the whole swarm alighted on a jasmine bush, leaving behind 8/9 of the swarm. And only one little, bee circled about a lotus for it was attracted by the buzzing of a sister bee that was so careless as to fall into the trap of the fragrant flower. How many bees were there in the swarm?

SOLUTION

If we denote the desired number of bees in the swarm by *x*, the equation takes the form

$$\sqrt{(x/2)} + (8x/9) + 2 = x.$$

This can be simplified by introducing an auxiliary unknown:

$$y = \sqrt{(x/2)}$$

Then $x = 2y^2$ and we obtain

$$y + (16y^2/9) + 2 = 2y^2 \text{ or } 2y^2 - 9y - 18 = 0.$$

Solving this equation, we get two values for y:

$$y_1 = 6, y_2 = -3/2$$

The corresponding values for x are

$$x_1 = 72, x_2 = 4.5.$$

Since the number of bees can only be whole and positive, only the first root satisfies the problem: the swarm consisted of 72 bees. Let us check this:

$$\sqrt{(72/2)} + (8/9)\,72 + 2 = 6 + 64 + 2 = 72.$$

A Troop of Monkeys

PROBLEM

Here is another Hindu problem that appears as a poem in a marvelous little book called *Who Invented Algebra?* by V. I. Lebedev. Rendered into English, it goes like this:

> Two little bands of monkeys at play.
>
> An eighth of them squared were jabbering wildly in a thicket
>
> When twelve shouted loudly with glee!
>
> Tell me: How many were there altogether in the thicket?

SOLUTION

If the total number of monkeys in the troop is x, then

$$(x/8)^2 + 12 = x$$

and from this

$$x_1 = 48, x_2 = 16$$

The problem has two positive solutions: there could be either 48 monkeys altogether or 16. Both answers fully satisfy the problem.

Farsighted Equations

In the problems we have so far considered, the two roots of each equation were made use of in different ways depending on the conditions of the problem. In the first case we dropped the negative root because it contradicted the sense of the problem, in the second case we discarded the fractional and negative root, and in the third problem, on the contrary, we made use of both roots. The existence of a second solution is often quite a surprise not only for the one working the problem but also for the one who thought it up. What follows is a problem in which the equation turned out to be more farsighted than the one who posed it.

A ball is thrown upwards at a speed of 25 meters a second. In how many seconds will it reach 20 meters above the ground?

SOLUTION

For bodies thrown upwards in the absence of air resistance, mechanics has established the following relationship between the height (h) it reaches above the ground, the initial velocity (v), the acceleration of gravity (g) and the time (t):

$$h = vt - (gt^2/2)$$

We disregard air resistance in this case because it is very slight in the case of small speeds. To further simplify calculations, we take g equal to 10 meters (instead of 9.8 meters, which is an error of only 2%). Substituting into our formula the values of h, v and g, we get the equation

$$20 = 25t - (10t^2/2)$$

which can be simplified to

$$t^2 - 5t + 4 = 0.$$

Solving this equation, we obtain

$$t_1 = 1 \text{ and } t_2 = 4.$$

The ball will be at a height of 20 meters twice: after a lapse of 1 second and after 4 seconds.

This appears to be improbable and so without giving much thought to the matter we discard the second solution. But that is a mistake! The second solution is quite meaningful, for the ball did indeed reach a height of 20 meters twice. First when it went up, and a second time on the way down. It can easily be figured out that with an initial velocity of 25 meters a second the ball will spend 2.5 seconds on its upward leg reaching a height of 31.25 meters. After 1 second it will reach 20 meters, but will go on upwards another 1.5 seconds. Then it will take the same amount of time to drop back to the 20-metres level and, a second later, will reach the ground.

Euler's Problem

Stendhal, in his autobiography, relates the following about his days of schooling:

"The mathematics teacher had a book by Euler and there I found his problem on the number of eggs that a peasant woman was carrying to market... . This was a revelation to me. I realized what it meant to use the tool called algebra. But, the devil take it, nobody had ever told me about this thing... ."

Here is the problem from Euler's *Introduction to Algebra* that so strongly impressed the young Stendhal.

Two peasant women together took 100 eggs to market, one had more than the other. Both sold them for the same sum of money. The first then said to the second: "If I had had your eggs, I would have earned 15 kreuzers," to which the Second replied: "If I had had your eggs, I would have earned 6 - 2 kreuzers." How many eggs did each have to begin with?

SOLUTION

Suppose the first peasant had x eggs and the second 100 - x. If the first had had 100 - x eggs, she would have earned 15 kreuzers. This means the first woman sold her eggs at

$$15/(100 - x)$$

apiece.

In the same way we find that the second peasant sold her eggs at

$$6\ 2/3 : x = 20/3x$$

apiece.

We can now determine the actual earnings of each peasant woman:

$$\text{First: } x \cdot 15/(100 - x) = 15\,x/(100 - x)$$

$$\text{Second: } (100 - x) \cdot 20/3x = 20\,(100 - x)/3x$$

Since they both earned the same amount it follows that

$$15x\ /\ (100 - x) = 20\,(100 - x)/3x$$

Simplifying we get

$$x^2 + 160x - 8000 = 0$$

and from this,

$$x_1 = 40, x_2 = -200.$$

Here the negative root is meaningless, and the problem has only one solution: the first peasant woman brought 40 eggs to market and the second, consequently, 60.

This problem can be solved faster, but it requires a good deal of insight and is harder to hit upon.

Suppose the second peasant had k times the number of eggs of the first. Since they earned the same money, the first peasant sold her eggs at a price k times that of the second. If they had exchanged their goods before selling them, the first peasant would have k times as many eggs as the second and would have sold them at k times the price. Which means she would have earned k^2 more money than the second peasant. And so we have:

$$k^2 = 15:6\ 2/3 = 45/20 = 9/4$$

whence

$$k = 3/2$$

It now remains to divide the 100 eggs in the ratio of 3 to 2. This tells us immediately that the first peasant woman had 40 eggs and the second 60 eggs.

Loudspeakers

PROBLEM

Thirteen loudspeakers are set up on a square in two groups: there are 4 in one group and 9 in the other. The two groups are separated by a distance of 50 meters. The question is: where should a person stand for the loudness of the sound from both groups to be the same?

SOLUTION

If we denote the distance of the desired point from the smaller group by x, then its distance from the larger group will be 50 - x (Fig. 16). Knowing that the sound intensity falls off with the square of the distance, we have the equation

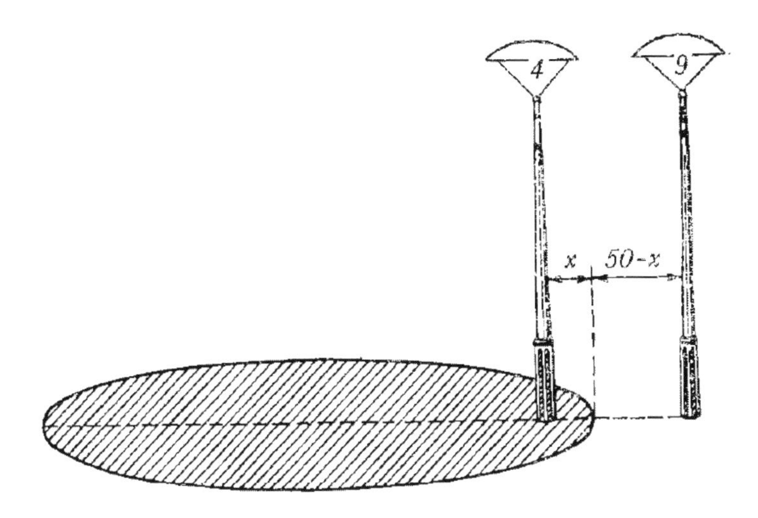

Fig. 16

$$4/9 = x^2 / (50 - x)^2$$

which, simplified, looks like this:

$$x2 + 80x - 2000 = 0.$$

Solving the equation we get two roots:

$$x_1 = 20,$$

$$x_2 = -100,$$

The positive root answers the question at once: the point of equal audibility is located 20 meters from the group made up of four loudspeakers and, consequently, 30 meters from the group of nine.

What does the negative root of the equation signify? Has it any meaning at all?

Yes, it does. The minus sign means that the second point of equal audibility lies in the *opposite* direction to that which is taken as the positive direction when we set up the equation.

If we lay off 100 meters from the location of the four loudspeakers in the required direction, we find the point where the sound from both groups of loudspeakers comes in with equal intensity. This point lies at a distance of

400 meters + 50 meters = 150 meters from the group of nine loudspeakers.

To summarize, then, we have found two points of equal audibility (that is, from among the points lying on a straight line joining the sources of sound). There are no other such points on that line, but outside the line there are. It can be proved that the set of all points satisfying our problem constitutes a circle drawn through the two points that were found as through the endpoints of the diameter. It will be seen that this circle bounds a rather extensive area (cross-hatched in the drawing) inside which the audibility of the group of four loudspeakers is greater than that of the group of nine, and outside this circle the situation is just the opposite.

The Algebra of a Lunar Voyage

In the same manner that we found points of equal audibility in a system of two groups of loudspeakers, we can find the points of equal attraction of a spaceship in flight between two celestial bodies: the earth and the moon. Let us try to find these points.

By Newton's law, the force of mutual attraction of two bodies is directly proportional to the product of the masses of the bodies and is inversely proportional to the square of the distance between them. If the mass of the earth is M and the distance of the spaceship from it is x, then the force with which the earth pulls each gram of mass of the spaceship is given as

$$Mk/x^2$$

where k is the force of mutual attraction of one gram by one gram at a distance of one centimeter.

The force with which the moon attracts each gram of the spaceship at the same point is

$$Mk/(l-x)^2$$

where m is the mass of the moon and 1 is its distance from the earth (the spaceship is assumed to be located between the earth and the moon on the straight line connecting their centers). The problem requires that

$$Mk/x^2 = Mk/(l-x)^2$$

Or

$$M/m = x^2 / (l^2 - 2lx + x^2)$$

The ratio M/m is known from astronomy to be roughly equal to 81.5. Substituting it into this equation gives us

$$x^2 / (l^2 - 2lx + x^2) = 81.5$$

and from this we get

$$80.5\, x^2 - 163.0\, lx + 81.5\, l^2 = 0.$$

Solving the equation for x, we obtain

$$x_1 = 0.9\, l,\ x_2 = 1.12\, l$$

As in the problem of the loudspeakers, we conclude that there are two desired points on the earth-moon line where the spaceship must be identically attracted by both celestial bodies: one point lying at 0.9 the distance between them reckoning from the centre of the earth, and the other at 1.12 the same distance. Since the distance l between the centers of the earth and the moon is approximately equal to 384 000 km, one of the sought-for points will be 346 000 km from the centre of the earth, and the other will be at 430 000 km.

Now we know (see the preceding problem) that all points of a circle passing through the two points just found, taking them as the endpoints of a diameter, have this property. If we rotate this circle about the line joining

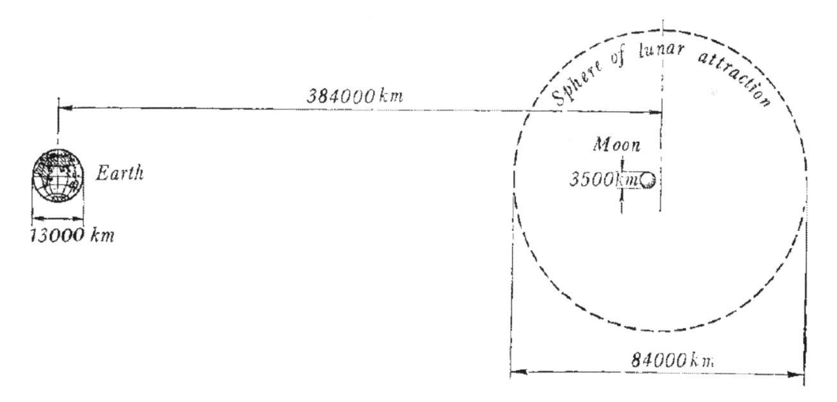

Fig. 17

the centers of the earth and the moon, it will describe a spherical surface, all points of which will satisfy the requirements of our problem.

The diameter of this sphere, called the sphere of attraction (Fig. 17) of the moon, is equal to

$$1.12\, l - 0.9\, l = 0.22\, l \approx 84\,000 \text{ km}.$$

There is a rather widespread erroneous opinion that if one wants to reach the moon in a spaceship it is sufficient to reach its sphere of attraction. At first glance it would appear that if a spaceship enters the sphere of attraction (and if, besides, its velocity is not considerable), then it must inevitably fall onto the moon since the force of lunar attraction in this region overcomes the earth's gravity, if this were the case, the problem of flights to the moon would be much simpler, for it would not be necessary to aim at the moon the diameter of which appears in the sky at an angle of 1/2°, but at a sphere of diameter 84 000 km, whose angular dimensions are equal to 12°.

However, it is easy to show that this reasoning is wrong.

Suppose a spaceship launched from the earth is continuously losing speed due to the earth's gravitational attraction and reaches the sphere of attraction of the moon with zero velocity. Will it fall onto the moon? Not at all!

First, even within the sphere of attraction of the moon, terrestrial gravity continues to be felt. That is why the force of lunar attraction off the earth-moon line will not merely overcome the force of attraction of the earth, but will combine with that force via the parallelogram rule and will yield a resultant force that is not at all directed towards the moon (it is only on the earth-moon line that this resultant force is directed towards the centre of the moon).

Second, and this is most important, the moon itself is not a fixed target, and if we want to know how a spaceship will move with respect to the moon (that is, whether it will reach the lunar surface or not), we have to take into account the velocity of the spaceship relative to the moon. Now this velocity is not at all equal to zero, since the moon itself is in motion about the earth with a velocity of 1 km/s. For this reason, the speed of the spaceship relative to the moon is too high for the moon to be aide to attract the spaceship or at least hold it within its sphere of attraction as an artificial satellite.

Actually, lunar attraction begins to exert an appreciable effect on the motion of a spaceship some time before the ship comes close to the sphere of attraction of the moon. In celestial ballistics, the rule is to take lunar gravity into account from the time the ship comes within the so-called *sphere of action of the moon* (66 000 km in radius). Then one can consider the motion of a spaceship relative to the moon and totally disregard the earth's gravity, but close account must be made of the velocity (relative to the moon) with which the spaceship enters the sphere of action. It is natural, therefore, that the spaceship has to be sent to the moon along a flight path that ensures that the velocity (relative to the moon) of entry into the sphere of action is directed at the moon. For this to happen, the sphere of lunar action must run into the spaceship as it is moving to an encounter across its path of travel. So we see that hitting the moon is no easy job, much less so than hitting a sphere 84 000 km in diameter.

Hard Problem

In 1895 N. Bogdanov-Belsky painted a picture called *A Hard Problem* (see Fig. 18) and many of those who have seen it most likely skipped over the problem itself, though it is well worth looking into. The idea behind the picture is a problem in mental arithmetic, to be solved at a glance:

$$(10^2 + 11^2 + 12^2 + 13^2 + 14^2) / 365 = ?$$

The problem really isn't easy. But the pupils of the teacher portrayed in the picture -and the portrait is an accurate one of S. A. Rachinsky, professor of natural science, who left the university to become an ordinary schoolteacher in the village- coped with it. In his school this talented teacher cultivated habits of mental arithmetic based on a marvelous handling of the properties of numbers. The numbers 10, 11, 12, 13, and 14 have the curious peculiarity that $10^2 + 11^2 + 12^2 = 13^2 + 14^2$.

Since $100 + 121 + 144 = 365$, it is easy to work out mentally that the expression given in the picture is equal to 2.

Algebra enables us to pose the problem of this interesting peculiarity of a number series on a broader basis: is this the only series of five consecutive numbers, the sum of the squares of the first three of which is equal to the sum of the squares of the last two?

SOLUTION

Fig. 18

Denoting the first of the desired numbers by x, we get the equation

$$x^2 + (x + 1)^2 + (x + 2)^2 = (x + 3)^2 + (x + 4)^2.$$

It is more convenient however to denote by x the *second* one of the sought-for numbers instead of the first. The equation then takes on a simpler aspect:

$$(x - 1)^2 + x^2 + (x + 1)^2 = (x + 2)^2 + (x + 3)^2.$$

Removing brackets and simplifying, we obtain

$$x^2 - 10x - 11 = 0,$$

whence

$$x = 5 \pm \sqrt{(25 + 11)},\ x_1 = 11,\ x_2 = -1.$$

Thus there are two sequences of numbers with the required property: the Rachinsky sequence

$$10, 11, 12, 13, 14$$

and the sequence

$$-2, -1, 0, 1, 2.$$

True enough,

$$(-2)^2 + (-1)^2 + 0^2 = 1^2 + 2^2.$$

Finding Numbers

PROBLEM

Find three successive numbers that have the property that the square of the middle number is greater by unity than the product of the other two numbers.

SOLUTION

If the first of the sought-for numbers is x, then the equation we can set up looks like this:

$$(x + 1)^2 = x(x + 2) + 1.$$

Removing brackets, we get the equation

$$x^2 + 2x + 1 = x^2 + 2x + 1.$$

But we can't find x. This means that we have an identity, and it holds true for all values of the letters involved and not only for certain values, as in the case of an ordinary equation. And so any three numbers taken in succession possess the required property. Indeed, take any three numbers at random,

$$17, 18, 19,$$

and we see that

$$182 - 17.19 = 324 - 323 = 1.$$

The necessity of such a relation is even more evident if we use x to denote the second number. We then get

$$x2 - 1 = (x + 1)(x - 1)$$

which is an obvious identity.

Chapter Seven

LARGEST AND SMALLEST VALUES

The problems in this chapter have to do with a very exciting thing: the seeking of maximum (largest) and minimum (smallest) values of some quantity. They may be solved in a variety of ways, one of which we give below.

In his paper entitled *The Drawing of Geographical Maps*, the eminent Russian mathematician P. L. Chebyshev wrote that those methods of science are of particular value that permit solving a problem that is common to the whole range of human activity: how to arrange the facilities at one's disposal so as to attain the greatest possible advantage.

Two Trains

PROBLEM

Two railway lines intersect at right angles. Two trains are racing at the same time to the intersection, one having left a station 40 km from the intersection, the other from a station 50 km from the intersection. The first train is doing 800 meters per minute, the second, 600 meters per minute.

In how many minutes after start will the locomotives be separated by the shortest distance? Find that distance.

SOLUTION

Let us make a diagram of the movements of the trains. Let the straight lines AB and CD be the intersecting lines (Fig. 19). Station B is 40 km from the point of intersection 0, station D is 50 km from it. Suppose that after a lapse of x minutes the locomotives are separated by the smallest distance $MN = m$. The train that left B would by then have covered the distance $BM = 0.8x$, since it does 800 meters, or 0.8 km, a minute.

And so $OM = 40 - 0.8x$. In the same way we find that $ON = 50 - 0.6x$. By the theorem of Pythagoras,

$$MN = m = \sqrt{(OM^2 + ON^2)} = \sqrt{[(40 - 0.8x)^2 + (50 - 0.6x)^2]}$$

Squaring both sides of the equation, we get

$$m^2 = (40 - 0.8x)^2 + (50 - 0.6x)^2$$

Simplifying, we obtain

$$x^2 - 124x + 4100 - m2 = 0.$$

Solving this equation for x, we finally have

$$X= 62 \pm \sqrt{(m2 - 256)}.$$

Since x is the number of minutes and that number cannot be imaginary, it follows that $m^2 - 256$ must be a positive quantity or at least zero. The latter corresponds to the smallest possible value of m, and then $m^2 = 256$, or m = 16.

It is obvious that in cannot be loss than 16, for then x becomes imaginary. And if $m^2 - 256 = 0$, then x = 62.

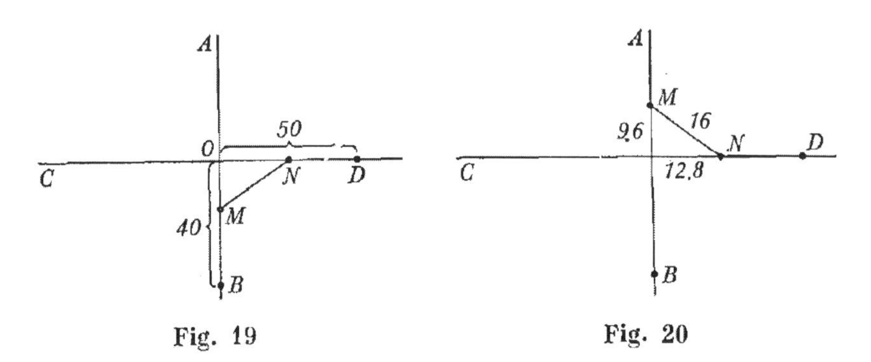

Fig. 19 Fig. 20

To summarize: the locomotives are closest to each other after 62 minutes and the separation will then be 16 km.

Let us now see how they stand at that moment. We compute the length of *OM* to be

$$40 - 62 \cdot 0.8 = -9.6.$$

The minus sign means that the locomotive has passed the intersection by 9.6 km. And the distance ON is equal to

$$50 - 62 \cdot 0.6 = 12.8$$

which means the second locomotive is 12.8 km short of the intersection. This position of the locomotives is shown in Fig. 20. As we now see, it doesn't look at all like what we imagined before we began. The equation turned out to be very tolerant and despite the incorrect drawing gave us the proper answer. This tolerance is clearly due to the algebraic rules for signs.

Planning the Site of a Flag Station

PROBLEM

Twenty kilometers from a railway line is a hamlet B (Fig. 21). The problem is to select a site for construction of a flag station C so that travelling time from A to B via railway from A to C and highway from C to B is a minimum. The rate of travel by rail is 0.8 kilometer a minute and by highway 0.2 kilometer a minute.

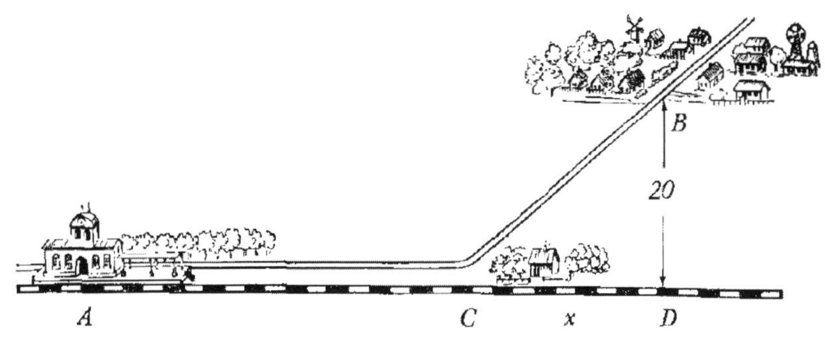

Fig. 21

SOLUTION

Denote by a the distance AD (from A to the foot of the perpendicular BD to AD) and by x the distance CD. Then AC = AD – CD = a – x and CB = $\sqrt{(CD^2 + BD^2)} = \sqrt{(x^2 + 20^2)}$. The time during which a train covers the distance AC is

$$AC/0.8 = (a - x)/0.8$$

The travelling time by highway from C to B is

$$CB/0.2 = \sqrt{(x^2 + 20^2)}/0.2$$

The time required to get from A to B is

$$(a - x)/0.8 + \sqrt{(x^2 + 20^2)}/0.2$$

This sum, which we denote by m, must be the smallest possible.

The equation

$$(a - x)/0.8 + \sqrt{(x^2 + 20^2)}/0.2 = m$$

can be cast in the form

$$-x/0.8 + \sqrt{(x^2 + 20^2)}/0.2 = m - a/0.8$$

Multiplying by 0.8, we get

$$-x + 4\sqrt{(x^2 + 20^2)} = 0.8m - a.$$

Denoting 0.8m - *a* by *k* and getting rid of the radical sign, we obtain the quadratic equation

$$15x^2 - 2kx + 6400 - k^2 = 0,$$

Whence

$$x = (k \pm \sqrt{(16k^2 - 96000)})/15.$$

Since k = 0.8m - a, it follows that k attains a minimum when m is a minimum and conversely. But for x to be real, $16k^2$ must be al, least equal to 96 000. Hence that figure is the smallest we can have for $16k^2$. Therefore, m becomes least when

$$16k^2 = 96000$$

and so we see that

$$k = \sqrt{6000}$$

and, consequently,

$$x = (k \pm 0)/\ 15 = \sqrt{6000}/15 \approx 5.16$$

To summarize, then, no matter how long a = AD is, the flag station must be located at about 5 km from the point D.

Quite naturally, our solution is meaningful only for cases where $x < a$, because when we set up the equation we regarded the expression a - x as being positive.

If $x = a \approx 5.16$, then there is no need to build a flag station at all and the highway will have to be built straight to the main station. And. the same goes for cases where the distance a is shorter than 5.16 km.

This time, it turns out, we have more insight than the equation. If we were to blindly follow the equation, we would have to construct a flag station right after the station, which would be nonsense: in that case $x > a$ and therefore the time

$$(a - x)/0.8$$

during which we travel by rail is negative. This is an instructive example that shows caution is necessary when handling mathematical tools and interpreting results. One must bear in mind that the final figures may be meaningless if the premises on which the use of those tools is based are not properly taken into account.

An Optimal Highway

PROBLEM

A town A is located on a river and we have to send freight to B, a town located a kilometers downstream and d kilometers from the river (Fig. 22). The problem is to locate a highway between B and the river so that transportation of goods from A to B is cheapest, taking into consideration that the transport cost of a ton-kilometer on the river is half that by highway.

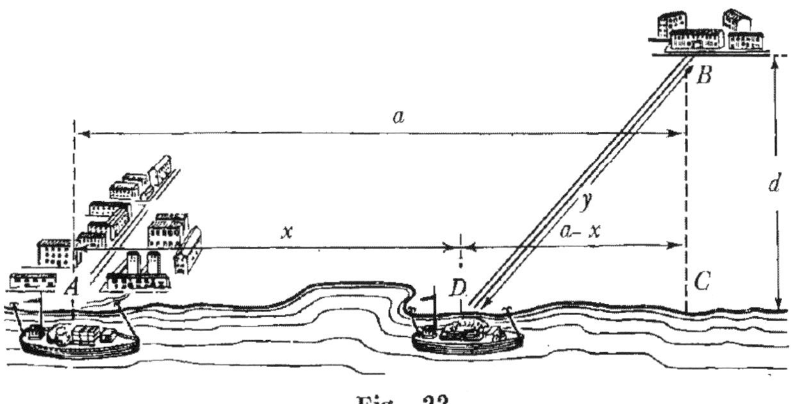

Fig. 22

SOLUTION

We denote the distance AD by x and DB by y: it is given that AC is equal to a and BC is equal to d.

Since highway transport is twice as expensive as river transport, the sum

$$x + 2y$$

must be a minimum in accordance with the requirements of the problem. We denote this minimum value by m and we have the equation

$$x + 2y$$

must be a minimum in accordance with the requirements of the problem. We denote the minimum value by m and we have the equation

$$x + 2y = m$$

But x = a - DC and ; and so our equation becomes

$$a - \ + 2y = m$$

or, after getting rid of the radical sign,

$$3y^2 - 4(m - a)y + (m - a)^2 + d^2 = 0.$$

Solving it for y, we get

$$y = 2/3(m-a) \pm$$

For y to be real, $(m - a)^2$ must be at least $3d^2$. The smallest value of $(m - a)^2$ is equal to $3d^2$ and then

$$m - a = d\sqrt{3},\ \ y = [2(m - a) + 0]/3 = 2d\sqrt{3}/3;$$

sin (BDC) = d: y, or

$$\sin(BDC) = d/y = d : (2d\sqrt{3}/3) = \sqrt{3}/2$$

Now, an angle whose sine is equal to $\sqrt{3}/2$ is 600. This means the highway must be sited at an angle of 60° to the river, no matter what the distance AC.

Here again we have the same peculiarity encountered in the preceding problem. The solution is meaningful only under a certain condition. If the point is located so that the highway built at an angle of 60° to the river passes on the other side of the town A, then the solution cannot be applied; in that case the highway should be built directly between B and the town A, dispensing with river transport.

When Is the Product a Maximum?

The solution of a large number of maximum and minimum problems (that is, seeking the largest and smallest values of some variable quantity) is neatly handled by an algebraic theorem that we now introduce. We reason as follows.

How can a given umber be partitioned into two parts so that the product of the parts is a maximum?

SOLUTION

Suppose the given number is a. Then the parts into which a is partitioned may be expressed as

$$a/2 + x \text{ and } a/2 - x$$

The number x shows by what amount these parts differ from a/2. The product of both parts is equal to

$$(a/2 + x)\ (a/2 - x) = a^2/4 - x^2$$

The product of the parts will clearly increase as x diminishes, that is to say, as the difference between the parts decreases.

And the product will be greatest when $x = 0$, which is the case when both parts are equal to a/2.

To summarize: the number must be halved -the product of two numbers whose sum is invariable will be a maximum when the numbers are equal.

Now let us examine the question of three numbers.

How do we partition a given number into three parts so that the product of the three parts is a maximum?

SOLUTION

We proceed on the basis of the foregoing problem. Suppose a number a is partitioned into three parts. To start with, we assume that no part is equal to a/3. Then there will be a largest part greater than a/3 (all three of them cannot be less than); we express it as

$$a/3 + x$$

There also will be a smallest part, less than a/3, which we express as

$$a/3 - y$$

The numbers x and y are positive. The third part will obviously be

$$a/3 + y - x$$

The numbers $a/3$ and $a/3 + x - y$ have the same sum as the first two parts of the number *a,* while the difference between them, or $x - y$, is less than the difference between the first two parts, which difference is equal to $x + y$. As we know from the solution of the preceding problem, it follows that the product

$$a/3\ (a/3 + x - y)$$

exceeds the product of the first two parts of a.

Thus, if we replace the first two parts of the number a by the numbers

$$a/3 \text{ and } a/3 + x - y$$

and leave the third part unchanged, then the product will increase.

Now let one of the parts be equal to $a/3$. Then the other two will look like this:

$$a/3 + x \text{ and } a/3 - z.$$

If we make the last two parts equal to $a/3$ (their sum does not change of course), then the product will again increase and become equal to

$$a/3.a/3.a/3 = a^3/\ 27$$

To summarize: if the number *a* is partitioned into three *a3* unequal parts, then the product of the parts is less than that is to say, than the product of three *equal* factors forming a sum equal to *a.*

In similar fashion we can prove this theorem for four factors, for five factors and so forth.

Now let us consider a more general case.

The problem is to find values of x and y such that the expression $x^p y^q$ is largest if $x + y = a$.

SOLUTION

For what value of x does the expression

$$x^p(a - x)^q$$

reach a maximum?

Multiply this expression by the number $1/p^pq^q$ to get a new expression

$$(x/p)^q((a - x)/q)^q$$

which will obviously attain a maximum value whenever the original expression does.

Let us represent the expression thus obtained as

$$(x/p)\cdot(x/p)\ldots\ldots(x/p)\cdot((a - x)/q)\ ((a - x)/q)\ldots((a - x)/q)$$

Where the number of (x/p) terms is p and the number of ((a - x)/q) terms is q.

The sum of all factors of this expression is equal to

$$x/p + x/p + x/p + \ldots.\ ((a - x)/q) + ((a - x)/q) + \ldots$$

Where the number of (x/p) terms is p and the number of ((a - x)/q) terms is q.

This sum is equal to:

$$(p\cdot x/p) + q\cdot(a - x)/q = x + a - x = a$$

that is, the quantity is a constant.

On the basis of what has been proved above we conclude that the product

$$x/p + x/p + x/p + \ldots.\ ((a - x)/q) + ((a - x)/q) + \ldots$$

reaches a maximum when all its separate factors are equal, that is, when

$$x/p = (a - x)/q$$

knowing that $a - x = y$, we obtain (after a simple manipulation) the proportion

$$x/y = p/q$$

Thus, the product x^py^q attains a maximum (given the constant sun $x + y$) when

$$x : y = p : q$$

In the same fashion we can prove that the products

$$x^p y^q z^r,\ x^p y^q z^r t^u \text{ and so forth}$$

(given the constant sums $x + y + z$, $x + y + z + t$ and so on) attain maximum values whenever

$$x : y : z = p:q:r,\ x:y:z:t = p:q:r:u \text{ and so on.}$$

When Is the Sum a Minimum?

The reader who would like to test himself in proving useful algebraic theorems can try his hand at proving the following propositions:

1. The sum of two numbers whose product is invariable becomes a minimum when the numbers are equal.

For example, take the product 36: 4 + 9 = 13, 3 + 12 = 15, 2 + 18 = 20, 1 + 36 = 37 and, finally, 6 + 6 = 12.

2. The sum of several numbers whose product is a constant becomes a minimum when the numbers are equal.

For example, take the product 216:3 + 12 + 6 = 21, 2 + 18 + 6 = 26, 9 + 6 + 4 = 19, yet 6 + 6 + 6 = 18.

The following are some cases where these theorems find practical application.

A Beam of Maximum Volume

PROBLEM

The problem is to saw out of a cylindrical log a rectangular beam of largest volume. Find the shape of the cross section it will have (Fig. 23).

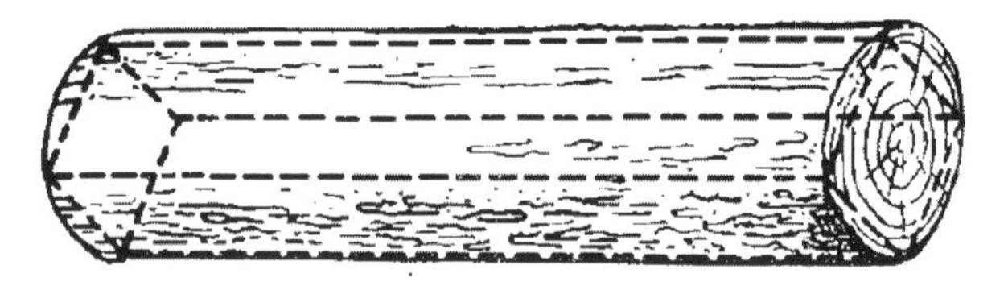

Fig. 23

SOLUTION

If the sides of the rectangular cross section are x and y, then by the Pythagorean theorem we have

$$x^2 + y^2 = d^2$$

where d is the diameter of the log. The volume of the beam is a maximum when the area of its cross section is a maximum, that is, when xy becomes a maximum. Now if xy is a maximum, then so is the product x^2y^2. Since the sum $x^2 + y^2$ is constant, it follows by what has been already proved that the product x^2y^2 is the largest possible one when

$$x^2 = y^2 \text{ or } x = y$$

Hence the cross section of the beam must be a *square*.

Two Plots of Land

PROBLEMS

1. What shape must a rectangular plot of land of a given area have for the length of fence bounding it to be a minimum?

2. What shape must a rectangular plot of land have for the area to be a maximum for a given length of fence?

SOLUTIONS

1. The shape of the rectangular plot depends on the relationship of its sides x and y. The area of a plot with sides x and. y is xy, and the length of the fence around the plot is $2x + 2y$. The fence will be of minimum length if $x + y$ attains a minimum.

For a constant product xy, the sum $x + y$ is a minimum when x equals y . Consequently, the sought-for rectangle is a square.

2. If x and y are the sides of a rectangle, then the length of the fence is 2x -1- 2y and the area is xy. The product will be a maximum at the same time that the product 4xy (or $2x + 2y$) is. Now the latter product,, given a constant sum of its factors, $2x + 2y$, becomes a maximum when $2x=2y$, which is when

the plot is a square.

To the familiar properties of a square that we learned in geometry we can add the following: of all rectangles the square has the smallest perimeter for a given area and the largest area for a given perimeter.

Making a kite

PROBLEM

We have a kite in the shape of a circular sector and it is required to change the shape so that it will have the largest possible area for the given perimeter. What will the shape of the sector be?

SOLUTION

Let us make more explicit the requirements of the problem: for what relationship between the length of the arc of the sector and its radius will the area attain a maximum for the given perimeter?

If the radius of the sector is equal to x and the arc is y then the perimeter l and the area S will be expressed as (see Fig. 24):

$$l = 2x + y$$

$$S = xy/2 = x\,(l - 2x)/2$$

The quantity S reaches a maximum for the same value of x as the product $2x\,(l - 2x)$, which is four times the area. Since the sum of the factors $2x + (l - 2x) = l$ is a constant quantity, their product is a maximum when $2x = l - 2x$, whence

$$x = l/4;$$

$$y = l - 2 \cdot l/4 = l/2$$

To summarize: for a given perimeter, the sector encloses the largest area when its radius is half the arc (or, the length of the arc is equal to the sum of the radii, or the length of the curvilinear portion of the perimeter is equal to the length of the broken line). The angle of the sector is approximately equal to 115°, or two radians. How the kite will fly is quite a different matter -something we won't go into.

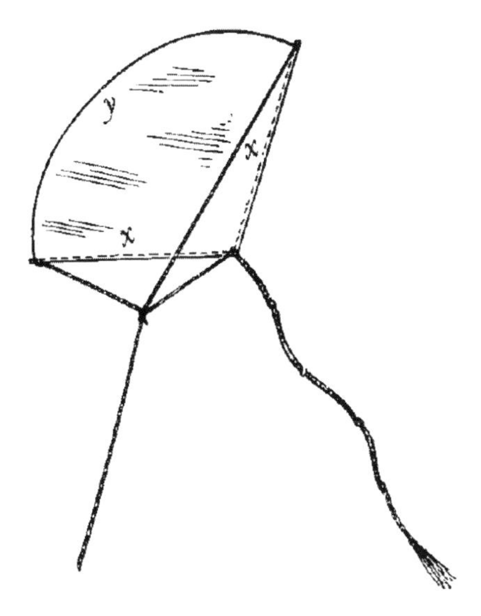

Fig. 24

Building a House

PROBLEM

Using one whole wall of a house that was partially destroyed, we wish to build a new house. The one remaining whole wall is 12 meters long. The area of the new house is to be 112 square meters. The conditions of work are as follows:

(1) repairing one linear meter of wall comes to 25% of laying down a new wall;

(2) dismantling one linear meter of the old wall and laying down a new wall using the material of the old wall will cost 50% of what constructing a linear meter of wall using new materials would come to.

What is the best way to make use of the old wall?

SOLUTION

Suppose x meters of the old wall are retained and the remaining 12 - x meters are dismantled so that the materials obtained are used again in the construction of a part of the wall of the new house (Fig. 25). if the cost of laying a linear meter of wall using new material is equal to a, then repairing x meters of the old wall will cost ax/4 ; erecting a portion of length 12 - x will cost a(12 - x)/2 the cost of the remaining portion of that wall will be a [y - (12 - x)] or a (y + x - 12); the cost of the third wall is ax, of the fourth wall, ay. Altogether, the work will cost

$$ax/4 + a(12 - x)/2 + a(y + x - 12) + ax + ay = a(7x + 8y)/4 - 6a$$

This expression is a minimum whenever the following sum is:

$$7x + 8y.$$

We know that the floor area xy of the house is 112; hence,

$$7x\,8y = 56\ 112.$$

Given a constant product, the sum $7x + 8y$ reaches a minimum when

$$7x = 8y$$

and so we have y = (7/8)x

Putting this expression for y into the equation xy = 112,

we got

$$(7/8)x^2 = 112,\ x = \sqrt{128} \approx 11.3.$$

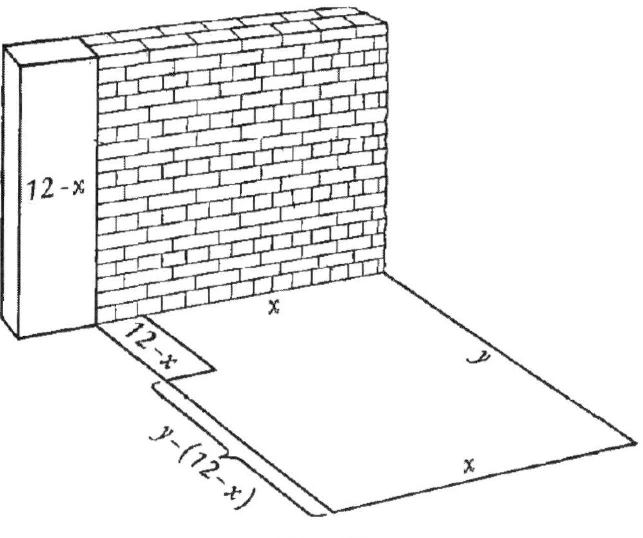

Fig. 25

And since the length of the old wall is 12 meters, we have to dismantle only 0.7 meters of that wall.

Fencing in a Lot

PROBLEM

The construction of a summer cottage is to be undertaken. The first step is to fence off the plot of land. There is enough material for l linear meters of fence. Also, a portion of an earlier built high wooden fence can be used for one side of the lot. Under these conditions, find a way to fence off a rectangular lot of largest possible area.

SOLUTION

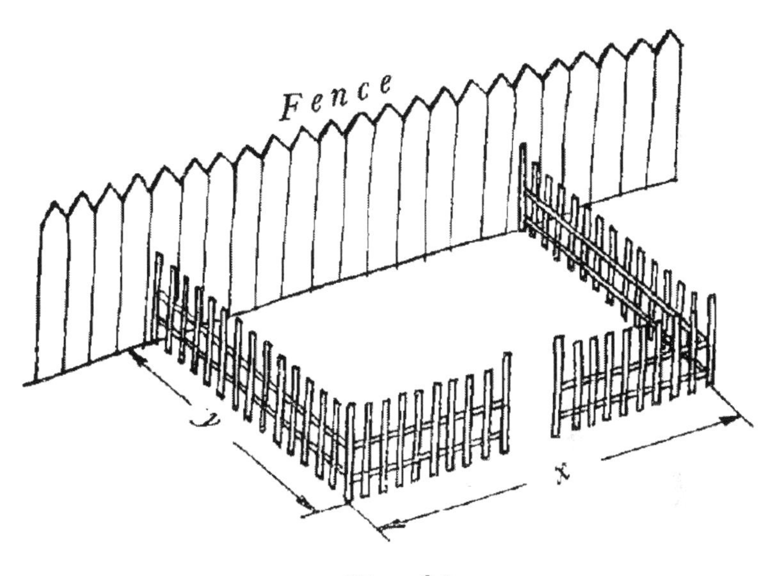

Fig. 26

Suppose the portion along the high fence (see Fig. 26) is x, while the width (that is, the dimension of the lot perpendicular to the high wooden fence) is equal to y. Then x + 2y meters of fencing is needed to enclose that portion, so that

$$x + 2y = l$$

The area of the lot is equal to

$$S = xy = y(1 - 2y)$$

It attains a maximum value at the same time that the quantity

$$2y(1 - 2y)$$

does. This is twice the area and is a product of two factors with a constant sum 1. And so to obtain the greatest area we must have

$$2y = 1 - 2y$$

Whence

$$y = 1/4, x = 1 - 2y = 1/2$$

In other words, $x = 2y$, or the length of the plot must be twice its width.

A Trough of Maximum Cross Section

PROBLEM

A rectangular sheet of metal (Fig. 27) is to be bent into the shape of a trough with cross section having the form of an equilateral trapezoid. This can be done in a variety of ways, as shown in Fig. 28. Of what width must the lateral strips be and at what angle are they to be bent for the cross section of the trough to have the maximum area (Fig. 29)?

SOLUTION

Let the width of the sheet be l. We denote the width of the bent side strips by x and the width of the bottom of the trough by y. Let us introduce another unknown z; its meaning is clear from the drawing in Fig. 30.

The area of the trapezoid that represents the cross section of the trough is

$$S = [(z + y + z) + y]/2 \sqrt{(x^2 - z^2)} = \sqrt{((y + z)^2 (x^2 - z^2))}$$

The problem reduces to determining those values of x, y, z for which S attains a maximum value; note that the sum $2x + y$ (that is, the width of the

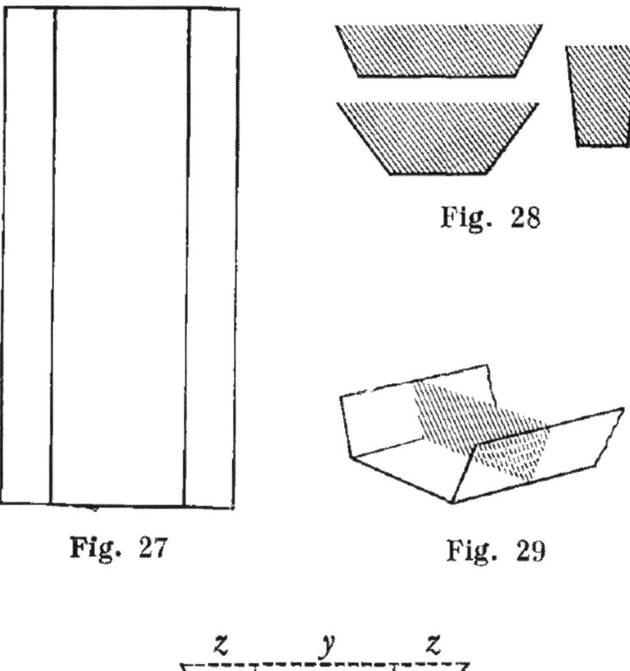

Fig. 27

Fig. 28

Fig. 29

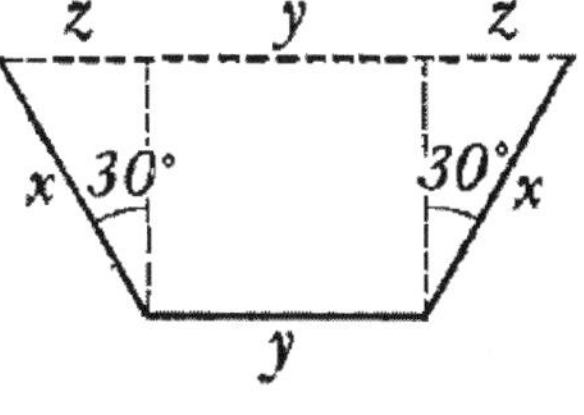

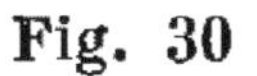

Fig. 30

sheet) retains a constant value l. After a few simplifying manipulations, we get

$$S^2 = (y + z)^2\ (x + z)\ (x - z)$$

The quantity S^2 is a maximum for the same values of x, y, z, as $3S^2$, and the latter can be expressed as a product:

$$(y + z)(y + z)(x + z)(3x - 3z)$$

The sum of these four factors

$$y + z + y + z + x + z + 3x - 3z = 2y + 4x = 2l,$$

is a constant. Therefore the product of our four factors is a maximum when they are equal to each other, or

$$y + z = x + z \text{ and } x + z = 3x - 3z$$

From the first equation we have

$$y = x,$$

and since $y + 2x = 1$, it follows that $x = y = 1/3$

From the second equation we find

$$z = x/2 = 1/6$$

Furthermore, since the leg z is equal to half the hypotenuse x (Fig. 30), the angle opposite that leg is equal to 30°, and the angle of inclination of the sides of the trough to the bottom is equal to 90° + 30° = 120°.

Summarizing, we see that the trough will have a maximum cross section when its faces are bent into the shape of three adjacent sides of a regular hexagon.

A Funnel of Maximum Capacity

PROBLEM

A circular tin disc is used to make the conical portion of a funnel. For this purpose (see Fig. 31), a sector is cut out and the remaining portion is twisted into a cone. How many degrees must there be in the arc of the cut-out sector so that the cone is of maximum capacity?

SOLUTION

Denote by x the length (in linear measure) of the arc of that portion of the circle that is twisted into a cone. Thus, the radius R of the tin disc will be the generatrix of the cone, and the base circle will be equal to x. We determine the radius r of the base of the cone from the equation

$$2\pi r = x, \text{ whence } r = x/2\pi$$

The altitude of the cone (by the Pythagoras theorem) is (Fig. 31):

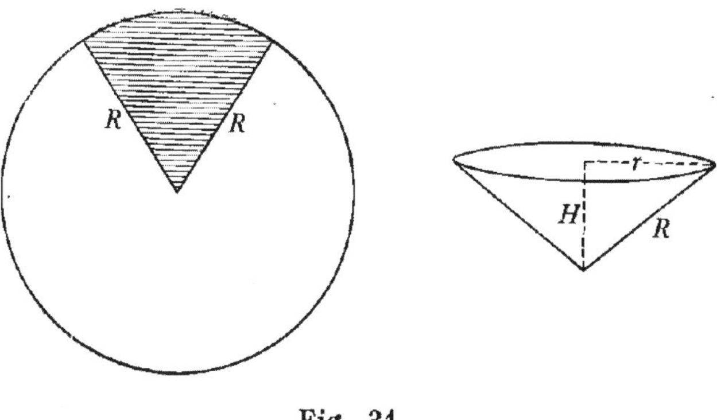

Fig. 31

$$H = \sqrt{(R^2 - r^2)} = \sqrt{(R^2 - x^2/4\pi^2)}$$

For the volume of the cone we have

$$V = (\pi/3)r^2 H = (\pi/3)(x/2\pi)^2\sqrt{(R^2 - x^2/4\pi^2)}$$

This expression becomes a maximum whenever the expression

$$(x/2\pi)^2\sqrt{(R^2 - x^2/4\pi^2)}$$

and its square

$$(x/2\pi)^4(R^2 - x^2/4\pi^2)$$

Since

$$(x/2\pi)^2 + R^2 - (x/2\pi)^2 = R^2$$

is a constant, it follows that the last product is a maximum for the value of x when

$$(x/2\pi)^2 : [R^2 - (x/2\pi)^2] = 2:1$$

And from this

$$(x/2\pi)^2 = 2R^2 - 2(x/2\pi)^2$$

$$3(x/2\pi)^2 = 2R^2 \text{ and } x = (2\pi/3)R\sqrt{6} \approx 5.15R$$

In degrees, the arc x ≈ 295° and, hence, the arc of the sector that was cut out must contain ≈ 65°.

The Brightest Illumination

PROBLEM

A candle is on a table (Fig. 32). At what height above the table must the flame be so as to best illuminate a coin lying on the table?

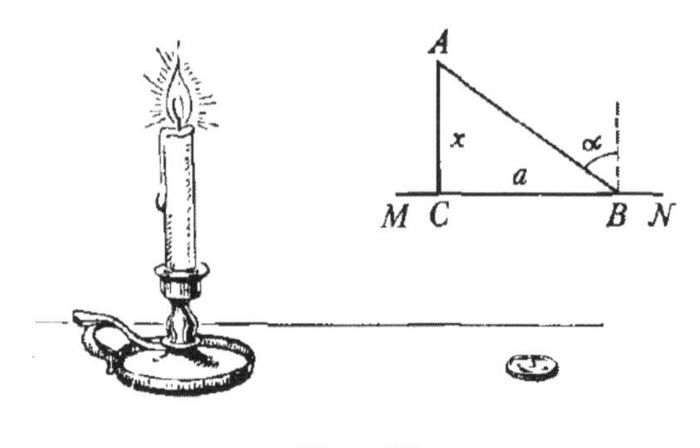

Fig. 32

SOLUTION

It might seem that the lower the flame the better. This is not so: in a low position of the flame cif the candle, the rays fall at a very sloping angle. But if the candle is high and the rays fall at a sharp angle, then the light source is too far away. The best illumination is clearly at some intermediate height of the flame above the table. We denote it by x (Fig. 32). And we use a to denote the distance BC of the coin B from the foot C of a perpendicular passing through the flame A. If the brightness of the flame is i, then the illumination of the coin is given, by ale laws of optics, as

$$(i/\sqrt{AB^2}) \cos \alpha = (i \cos \alpha)/(\sqrt{(a^2 + x^2)})^2 = (i \cos \alpha)/(a^2 + x^2)$$

Where α is the angle of incidence of a pencil of rays AB.

Since

$$\text{Cos } \alpha = \cos A = x/AB = x/\sqrt{(a^2 + x^2)}$$

it follows that the illumination is

$$[i/(a^2 + x^2)][x/\sqrt{(a^2 + x^2)}] = ix/(a^2 + x^2)^{3/2}$$

This expression reaches a maximum for the same value of x as its square, or

$$i^2x^2/(a^2 + x^2)^3$$

The factor i^2 is dropped since it is a constant; the remaining part of the expression is then transformed as follows:

$$x^2/(a^2 + x^2)^3 = [1/(a^2 + x^2)^2][1 - a^2/(a^2 + x^2)] = [1/(x^2 + a^2)]^2[1 - a^2/(x^2 + a^2)]$$

The transformed expression reaches a maximum together with the expression

$$[a^2/(x^2 + a^2)]^2[1 - (a^2/(x^2 + a^2))]$$

since the constant factor a^4 that was introduced does- not affect the value of x for which the product attains its maximum value. Noting that the sum of the first powers of these factors,

$$a^2/(a^2 + x^2) + [1 - (a^2/(a^2 + x^2))] = 1,$$

is a constant, we conclude that the product at hand is a maximum when

$$a^2/(x^2 + a^2) : (1 - a^2/(x^2 + a^2)) = 2:1$$

We have the equation

$$a^2 = 2x^2 + 2a^2 - 2a^2.$$

Solving this equation, we find that

$$x = a/\sqrt{2} \approx 0.71a.$$

The coin is best illuminated when the light source is at a height of 0.71 of the distance from the projection of the source to the coin. Knowing this relation helps in arranging proper lighting of one's work place.

Chapter Eight

PROGRESSIONS

The Most Ancient Problem Dealing with Progressions

PROBLEM

The oldest problem involving progressions is not that of awarding the inventor of chess -over two thousand years ago. There is a much older problem that has to do with dividing loaves of bread and is recorded in the famous Rhind papyrus of Egypt. This papyrus which was discovered by Rhind at the end of last century was written about 2000 years B.C. and is an excerpt from a still more ancient mathematical text that perhaps belongs to the third mil Eenium before our era. Included among the arithmetical, algebraic and geometric problems of that document is the following one which we give in a free translation:

Divide one hundred loaves of bread among five persons so that the second one receives as much more than the first as the third receives more than the second, and the fourth more than the third, and the fifth more than the fourth. Besides, the first two are to receive 7 times less than the three others. How much is to be given to each?

SOLUTION

Clearly, the loaves obtained by those participating in the division constitute an increasing arithmetic sequence (progression). Let the first term be x and the difference y. Then

the portion of the first man is x

the portion of the second man is $x+y$

the portion of the third man is $x+2y$

the portion of the fourth man is $x+3y$

the portion of the fifth man is $x+4y$

Using the conditions given in the problem, we set up the following two equations:

$$(x+y)+(x+2y)+(x+3y)+(x+4y)=100,$$

$$7[x + (x + y)] = (x + 2y) + (x + 3y) + (x + 4y).$$

After simplifications the first equation looks like this:

$$x + 2y = 20,$$

and the second like this:

$$11 x = 2y.$$

Solving this system of equations, we obtain

$$x = 1\ 2/3,\ y = 9\ 1/6$$

And so the loaves of bread are to be divided into the following portions:

$$1\ 2/3,\ 10\ 5/6,\ 20,\ 29\ 1/6,\ 38\ 1/3$$

Algebra on Squared Paper

Despite the nearly 50 centuries that this problem in progression has been around, it found its way into school only a relatively short while ago. Take the Russian textbook of Magnitsky published two hundred years ago that served as the standard school text for half a century; here, progressions are given, but there are no general formulas relating the quantities involved. For that reason, even the writer of the textbook himself found such problems hard. Yet it is so easy to derive the formula for the sum of the terms of an arithmetic progression in a simple and pictorial manner with the aid of squared paper. On such paper, any arithmetic progression can be depicted as a step-like figure. For example, the diagram ABDC in Fig. 33 depicts the progression

$$2, 5, 8, 11, 14.$$

In order to determine the sum of its terms, fill out the diagram to complete the rectangle ABGE. We then have two equal figures: ABDC and DGEC. The area of each describes the sum of the terms of our progression. Hence, the double sum of the progression is equal to the area of the rectangle ABGE, or

$$(AC + CE) \cdot AB.$$

But AC + CE gives the sum of the first and fifth terms of the progression; AB is the number of terms in the progression. Therefore, the double sum:

2S = (the sum of the extreme terms) x (the number of terms)

Or

S = (first term+ last term) x (number of terms)/2

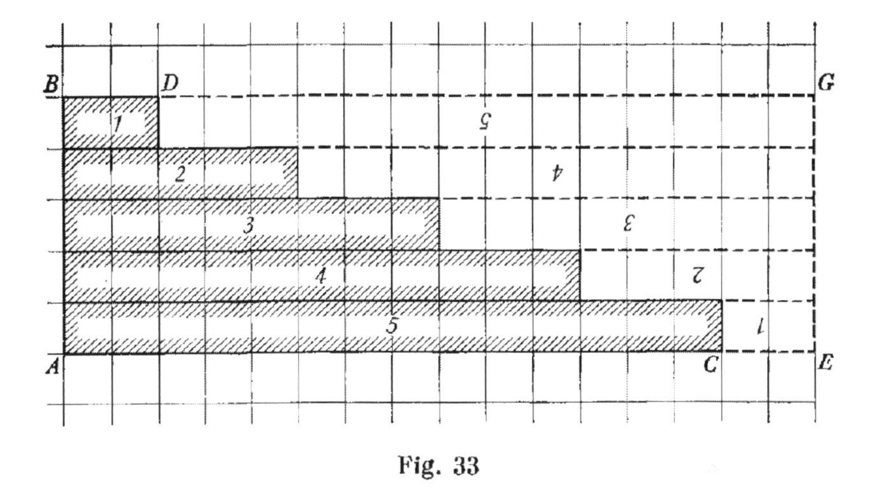

Fig. 33

Watering the Garden

PROBLEM

There are 30 rows in a garden, each row 16 meters in length and 2.5 meters in width. The gardener waters his garden by hauling pails of water from a well 14 meters from the edge of the garden, and then walks between the rows. One trip to the well suffices to water only one row.

What distance does the gardener cover in watering the whole garden? The start and finish are at the well.

SOLUTION

To water the first row, the gardener covers the distance 14 + 16 + 2.5 + 16 + 2.5 + 14 = 65 meters.

In watering the second row, he covers

$$14 + 2.5 + 16 + 2.5 + 16 + 2.5 + 2.5 + 14 = 65 + 5 = 70 \text{ meters.}$$

Each subsequent row requires 5 meters more to be covered than the preceding one. We then get the progression

$$65, 70, 75; \ldots; 65 + 5 \times 29.$$

The sum of the terms of the progression equals

$$(65 + 65 + 29.5)\ 30/2 = 4125 \text{ meters.}$$

In all, the gardener covers a distance of 4.125 km in watering his garden.

Feeding Chickens

PROBLEM

A certain amount of feed has been stored for 31 chickens, to be used at the rate of one decaliter (10 liters) per chicken per week. There was to be no change in the number of chickens. But since there vv-as a decrease of one chicken every week, the feed lasted twice as long as planned.

What was the original supply of feed and for what period was it planned?

SOLUTION

Suppose x decalitres of feed was stored for y weeks. Since it was calculated for 31 chickens at the rate of 1 decalitre per chicken per week, we have

$$x = 31y$$

During the first week, 31 decalitres was used up, during the second 30, during the third 29, and so on up to the last week of twice the originally planned period. The amount of feed consumed in the last week came to

$$(31 - 2y + 1) \text{ decalitres.}$$[1]

The total supply thus came to

$$x = 31y = 31 + 30 + 29 + \ldots + (31 - 2y + 1).$$

-1- The fourteen-place logarithms of Briggs are, incidentally, available only for the numbers from 1 to 20 000 and from 90 000 to 101 000.

The sum of 2y terms of the progression, the first term of which is 31 and the-last term is, 31 - 2y + 1, is equal to

$$31y = (31 + 31 - 2y + 1)\ 2y/\ 2 = (63 - 2y)\ y.$$

Since y cannot be zero, we have every right to divide through by that factor. And we get

$$31 = 63 - 2y \text{ and } y = 16,$$

Whence

$$x = 31y = 496.$$

The supply that was put in came to 496 decalitres of feed calculated to last 16 weeks.

A Team of Diggers

PROBLEM

The senior-class boys at a school took upon themselves the digging of a ditch in the school yard and organized a team of diggers. If the team had worked all at once, the ditch would have been ready in 24 hours. Actually, however, only one boy began. After a time, another boy joined him, and then -again after the same amount of time- a third boy joined in, then a fourth, and so on up to the last one. It was found that the first had worked 11 times longer than the last.

How much time did the last boy work?

SOLUTION

Suppose the last member of the team worked x hours. Then the first one worked 11x hours. Now if the number of diggers was y, then the total number of hours of work will be expressed as the sum of y terms of a decreasing progression, the first term of which is 11x and the last term is x,

$$(11x + x)\ y\ /\ 2 = 6\ xy$$

On the other hand it is known that a team of y boys working all together would be able to dig the ditch in 24 hours, which means 24y working hours is needed to complete the job.

Consequently,

$$6xy = 24y.$$

The number y cannot be zero and so we can cancel it out of the equation to get

$$6x = 24$$

and

$$x = 4.$$

This means the last one of the team to start working was busy 4 hours.

We have found the answer to the problem, but if we had been curious enough to ask how many participated in the work, we wouldn't have been able to say, despite the fact that that number entered into the equation (it was y). It is merely that there is not enough information in the hypothesis of the problem to get that answer.

Apples

PROBLEM

A man has an apple orchard and sells to his first customer half of all the apples plus half an apple; to the second customer he sells half of the rest plus half an apple; to the third, half of the remainder plus half an apple, and so on. To the seventh customer he sells half of what remains and another half-apple. And that is all he had. How many apples did the farmer start out with?

SOLUTION

Use x to denote the original number of apples; then the first customer received

$$(x/2) + (1/2) = (x+1)/2$$

the second customer,

$$1/2\ (x - (x+1)/2) + 1/2 = (x+1)/2^2$$

the third,

$$1/2\ (x - (x+1)/2 - (x+1)/4) + 1/2 = (x+1)/2^3$$

and the seventh customer bought

$$(x+1)/2^7$$

We thus have the following equation:

$$(x+1)/2 + (x+1)/2^2 + \ldots + (x+1)/2^7 = x$$

Or

$$(x+1)\ (1/2 + 1/2^2 + \ldots + 1/2^7) = x$$

Computing the sum of the terms of the geometric progress in the parentheses, we get

$$x/(x+1) = 1 - 1/2^7$$

And

$$x = 2^7 - 1 = 127.$$

Altogether there were 127 apples.

Buying a Horse

PROBLEM

In the old Russian arithmetic of Magnitsky we find an amusing problem that I give here in a translation into modern language.

Somebody sold a horse for 156 rubles. However the buyer, after all, decided not to take the horse and so returned it to its owner with the words:

"There is no point in my taking this horse for such a price because the animal is just not worth it."

Then the owner suggested the following.

"If the price of my horse is too high, then buy only the horseshoe nails, and then PH give you the horse free of charge. There are six nails in each horseshoe. For the first nail you give me only of a kopeck, for the second 1/2 , for the third 1 kopeck, and so on,"

Fig. 34

The client was mighty pleased to hear such a low price and, wishing to get the horse free of charge, agreed to the conditions of the owner, figuring that he would not have to pay more than 10 rubles altogether.

How much did the buyer lose in this deal?

SOLUTION

For 24 horseshoe nails he had to pay

$$1/4 + 1/2 + 1 + 2 + 2^2 + 2^3 + \dots + 2^{24-3}$$

kopecks. This is equal to

$$(2^{21} \cdot 2 - 1/4)/\ (2-1) = 2^{22} - 1/4 = 4194303\ ¾ \text{ kopecks}$$

or about 42 000 rubles. With a price like that, there is no need to worry about the horse going free of charge.

Paying for the Wounds of a Soldier

PROBLEM

Another old Russian textbook of mathematics that goes by the lengthy title *A Complete Course in Pure Mathematics Compiled by Cadet of Artillery and Special Mathematics Teacher Elim Voityakhovsky for the Benefit and Use of the Youth and Those Practising Mathematics* (1795) offers this problem:

A soldier was rewarded for each wound received in battle: for the first wound he got I kopeck, for the second 2 kopecks, for the third 4 kopecks, and so on. When the complete calculation was made, it was found that the soldier was rewarded the sum of 655 rubles and 35 kopecks. We want to know the number of wounds he received.

SOLUTION

We set up the equation

$$65\,535 = + 2 + 2^3 + 2^3 + \ldots + 2^{x-1}$$

Or

$$65\,535 = (2^{x-1}\,2 - 1)/(2 - 1) = 2^x - 1$$

and from this we find

$$65\,536 = 2^x \text{ and } x = 16.$$

This result can be easily found by running through the sequence.

With this generous system of payment, a soldier has to get 16 wounds and be still alive in order to reap his reward of 655 rubles and 35 kopecks.

Chapter Nine

THE SEVENTH MATHEMATICAL OPERATION

The Seventh Operation

We have already mentioned that the fifth operation -raising to a power- has two inverse operations. If

$$a^b = c$$

then finding a is one inverse (extraction of a root) and finding b is the other inverse: (taking logarithms). I am sure the reader has some basic knowledge of logarithms from his school studies. He would probably have no difficulty in figuring out the meaning of the following expression:

$$a^{\log_a (b)}$$

It is easy to see that if the base a of logarithms is raised to the power of the logarithm of the number b, then the result must be the number b.

Why were logarithms invented? To speed up and simplify calculations, naturally. The inventor of the first tables of logarithms, Napier, spoke of the impulse to his work, saying he had tried to the best of his ability to remove the difficulties and boredom of calculation, which ordinarily is so tiresome as to frighten away many from the study of mathematics.

Indeed, logarithms simplify and speed up calculations to a remarkable degree, to say nothing of the fact that they make it possible to perform operations that would otherwise be extremely difficult (extracting high-index roots).

With full justification, Laplace could write that the invention of logarithms, by reducing the amount of calculation from the work of months to that of days, actually doubled the life of astronomers. The great mathematician spoke of astronomers because it was their lot to carry out particularly complicated and arduous computations. But his words can justly be applied to everyone who has to deal with numerical computations.

Today we are used to logarithms and to the extent to which they simplify the computation process and so it is hard to imagine the wonder and excitement they caused when they first appeared. A contemporary of Napier's, Henry Briggs, who later came to fame through the invention of common (based on ten) logarithms, wrote the following in a letter after having read Napier's work: "Napper [Napier], lord of Markinston, 'lath set my [lead and hands a work with his new and admirable logarithms. I hope to see -him this summer, if it please God, for I never saw book which pleased me better, or made me more wonder." Briggs set out for Scotland to visit the inventor of logarithms. When they met, Briggs began: "My lord, I have undertaken this long journey purposely to see your person, and to know by what engine of wit or ingenuity you came first to think of this most excellent help in astronomy, viz, the logarithms; but, my lord, being by you found out, I wonder nobody found it out before, when now known it is so easy."

Rivals of Logarithms

Before the invention of logarithms, the requirements of speedy calculations gave rise to tables of a different kind in which the operation of multiplication is replaced by subtraction instead of addition. These tables are based on the identity

$$ab = (a+b)^2/4 - (a-b)^2/4$$

All one needs to do is open the brackets to see that the identity holds.

Having ready at hand the fourths of squares, one can find the product of two numbers without performing the multiplication but simply by subtracting the fourth of the square of the difference of the numbers from that of their sum. The same tables simplify squaring numbers and taking the square roots of numbers, and when combined with a table of reciprocals, they simplify the operation of division as well. Their advantage over logarithmic tables is that they yield exact, not approximate, results. On the other hand, however, they are inferior to logarithmic tables in a number of ways which are of more practical importance. Whereas tables of the fourths of squares permit multiplying only two numbers, logarithms enable one to find at once the product of any number of factors, and, what is more, to raise a number to any power and extract roots having arbitrary indices (integral and fractional). For example, it is impossible to compute compound interest with the aid of tables of fourths of squares.

Even so, tables of fourths of squares continued to be published after the appearance of a great variety of logarithmic tables. In 1856, a set of tables appeared in France entitled:

A table of. the squares of numbers from 1 to 1000 million with the aid of which one can find the exact product of numbers by an extremely simple procedure that is more convenient than by means of logarithms. Compiled by Alexander Cossar.

This very idea pops up time and again without their inventors realizing that it is an old one. I personally was approached by two inventors of similar tables and they were greatly surprised to learn that their invention was already over three hundred years old.

A different and younger rival of logarithms are the computation tables of most engineering reference works. These are combination tables with columns of squares, cubes, square roots, cube roots, reciprocals, circumferences and areas of circles for numbers from 2 to 1000. Such tables are very convenient for many engineering calculations but they are not always sufficient. Logarithmic tables have a far broader range of application.

The Evolution of Logarithmic Tables

Not so long ago Soviet schools used five-place logarithmic tables. They have now gone over to four-place tables because these are quite sufficient for engineering calculations. But for most practical needs, even three-place mantissas are suitable. The point is that only rarely do measurements involve more than three decimal places.

The idea that shorter mantissas would suffice was realized just recently. I can remember a time when we used hefty tomes of seven-place logarithms in school. They were later replaced by five-place tables after a stubborn struggle. But even the seven-place logarithms seemed to be an outrageous innovation when they appeared in 1794. The first common logarithms created by the energy of the London mathematician Henry Briggs (1624) were 14-place. A few years later they were supplanted by the ten-place tables of the Dutch mathematician Adrian Vlacq.

As we have seen, the evolution of practical logarithmic tables has been in the direction from mantissas with many decimal places to fewer and fewer decimal places, and the trend is still in progress today because many people do not realize the simple fact that the accuracy of computations cannot exceed that of the measurements.

Shortening the mantissas brings about two important practical results: (1) a much smaller volume of the tables and (2) a concomitant simplification in their use, which of course means faster calculations. Seven-place logarithms of numbers take up about 200 large-size pages, five-place logarithms take up only 30 pages of a much smaller size book; four-place logarithms are ten times shorter, occupying only two large-size pages, and three-place logarithms fit into a single page.

Now as to speed of calculation, it has been found that for instance a calculation via five-place tables requires only one third the time seven-place tables do.

Logarithmic-Table Champions

Whereas the computational needs of practical life and general engineering are quite satisfied with three- and four- place tables, the theoretician has need for tables with many more places than even the fourteen-place logarithms of Briggs. Generally speaking, in most cases a logarithm is an irrational number and cannot be exactly expressed by any number of digits: the logarithms of most numbers are given only approximately, no matter how many decimal places are taken -the larger the number of decimal places in the mantissa, the more exact the result. Even the Briggsian fourteen-place tables[1] prove insufficient for some scientific work. However, the researcher will never go unsatisfied, for among the 500 various types of logarithmic tables that have appeared since logarithms were invented there will definitely be one that can handle the job. Let us take, for example, the twenty-place logarithms of numbers from 2 to 1200 that were published in France by Callet (1795). For smaller ranges of numbers there are tables of logarithms with fantastic numbers of decimal places. There are giants that many mathematicians have never even suspected existed.

Here is a short list of the champion logarithms (they are all natural logarithms, not common logarithms):

48-place tables of Wolfram for the numbers up to 10 000;

61-place tables of Sharp;

102-place tables of Parkhurst;

and, finally, the logarithmic wonder of them all: the 260-place logarithms of Adams.

The last case, to be exact, is not a table but only the so-called natural logarithms of five numbers: 2, 3, 5, 7 and 10 and the 260-decimal conversion factor for converting them to common logarithms. But it is easy to see that

-1- Natural logarithms use the base 2.718... (instead of 10). They are discussed later on.

with the logarithms of these five numbers it is possible, via simple addition or multiplication, to obtain the logarithms of a multitude of composite numbers; for example, the logarithm of 12 is equal to the sum of the logarithms of 2, 2 and 3, and so forth.

Another logarithmic marvel is the slide rule ("wooden logarithms") but it has become so common a counting tool of the engineer as to be on the level of yesteryear's abacus among clerical workers. The slide rule is such a routine tool we are no longer amazed that while it operates on the principle of logarithms, the user need not even know what a logarithm is.

Logarithms on the Stage

One of the most amazing feats performed on the stage by professional calculators is the one where the lightning calculator offers to mentally work out the high-index roots of multidigit numbers. At home you arduously calculate the 31st power of some number and are prepared to knock out the calculating virtuoso with a 35-digit leviathan. At the right moment you ask the calculator:

"Try to find the 31st root of the following 35-digit number. Write it down as I dictate."

The calculator takes a piece of chalk and before you have opened your mouth with the first digit he has already written down the result: 13.

Without even knowing the number, he extracted the 31st root in his head and with lightning speed!

You sit flabbergasted, yet there is nothing at all supernatural in this feat. The secret is that there is only one number, namely 13, which, raised to the power of 31, yields a 35-digit result. Numbers less than 13 produce fewer than 35 digits, those greater than 13 generate more digits.

But how did the calculator know that? How did he find the number 13? Very simply, with the aid of logarithms, two-digit logarithms that he had learned by heart for the first 15 to 20 numbers. It is no trouble at all to memorize them, especially if one bears in mind that the logarithm of a composite number is equal to the sum of the logarithms of its prime factors. Knowing the logarithms of 2, 3 and 7 10

(recall that log 5 = log 10/2 = 1 - log 2), you already know the logarithms of the first ten numbers; for the second ten, one has to memorize the logarithms of another four numbers.

At any rate, the lightning calculator of the stage has in his memory the following table of two-digit logarithms:

Number	Logarithm	Number	Logarithm
2	0.30	11	1.04
3	0.48	12	1.08
4	0.60	13	1.11
5	0.70	14	1.15
6	0.78	15	1.18
7	0.85	16	1.20
8	0.90	17	1.23
9	0.95	18	1.26
		19	1.28

The mathematical trick that amazed you is this:

$$\log \sqrt[31]{(35 \text{ digits})} = \frac{34, \ldots}{31}$$

The desired logarithm can lie between

34/31 and 34.99/31 or between 1.09 and 1.13

In this interval we have the logarithm of only one whole number, namely 1.11, it is the logarithm of 13. That is exactly bow the startling result was found. True, to do all this mentally in a flash requires the training and skill and wit of the professional, but essentially it is quite simple, as you can see. Tricks of this kind are now within your grasp, if not mentally then at least on paper.

Suppose you are asked to find the 64th root of a 20-digit number.

Without even asking for the number you can state the result of the extraction: the root is 2.

Indeed , Consequently, it lies between 19/64 and 19.99/64, or between 0.29 and 0.32. There is only one such logarithm for a whole number: 0.30..., or the logarithm of the number 2.

You can even demolish your questioner by telling him what number he was about to propose: the famous chess number

$$2^{64} = 18\ 446\ 744\ 073\ 709\ 551\ 616.$$

Logarithms on a Stock-Raising Farm

PROBLEM

The amount of the so-called maintenance ration (that is the minimum feed required to maintain the needs of the body for heat emission, the functioning of the internal organs, the restoration of dying cells, and so forth)[2] is proportional to the surface area of the animal. Knowing this, determine the calorific value of maintenance feed for an ox weighing 420 kilograms if under the same conditions an ox weighing 630 kilograms requires 13 500 calories.

SOLUTION

To work out this practical problem of stock farming, we will need some geometry in addition to algebra. It is given that the desired calorific value x is proportional to the surface (s) of the ox, or

$$x/13500 = s/s_1$$

where s_1 is the surface area of an ox weighing 630 kg, Geometry says that the surface areas (s) of similar bodies are in the same ratio as the squares of their linear dimensions (1), and the volumes (hence, the weights) are in the same ratio as the cubes of their linear dimensions. Therefore

$s/s_1 = l^2/l_1^2$, $420/630 = l^3/l_1^3$ and, hence

$$\frac{l}{l_1} = \frac{\sqrt[3]{420}}{\sqrt[3]{630}}$$

From this we get

-2- In contrast to the "productive" ration, which is the part that goes to build up the animal for slaughter.

$$\frac{x}{13\,500} = \frac{\sqrt[3]{420^2}}{\sqrt[3]{630^2}} = \sqrt[3]{\left(\frac{420}{630}\right)^2} = \sqrt[3]{\left(\frac{2}{3}\right)^2},$$

$$x = 13\,500\sqrt[3]{\frac{4}{9}}.$$

Using a table of logarithms, we find that

$$x = 10\ 300$$

Which means the ox requires 10 300 calories.

Logarithms in Music

Musicians do not often take a liking to mathematics; most of them respect the science but prefer to stay away from it. Yet even those musicians win: do not verify "harmony by means of algebra" (like Pushkin's Salieri) come into contact with mathematics much more frequently than they even suspect; what is more, their contact is with such frightful things as logarithms.

I permit myself here a short quotation from an article by the later physicist Professor A. Eichenwald. It appeared in the *Russian Astronomical Calendar for 1919* and was entitled "On Large and Small Distances."

"A friend of mine from Gymnasium days liked to play the piano but detested mathematics. He spoke with a touch of scorn about music and mathematics having nothing whatsoever in common. 'True, Pythagoras found some kind of relationships between sound vibrations, but it is precisely the Pythagorean scale that turned out to be unsuitable for our music.'

"Imagine the surprise of my friend when I showed him that in running his hands over the keys of a modern piano he was actually playing on logarithms. Indeed, the steps of the tempered chromatic scale are not arranged at equal distances either with respect to the number of vibrations or with respect to the wavelengths of the appropriate sounds; they are the logarithms of these quantities. Only the base of the logarithms is 2 instead of 10, as commonly used.

"Suppose the note do of the lowest octave (we will call it the zero octave) is given as n vibrations per second. Then the note do of the first octave ,vill have 2n vibrations, that of the mth octave, $n \cdot 2^m$ vibrations, and so on. Let us denote all notes of the chromatic scale of a piano by the numbers p, assuming the fundamental tone do of each octave to be the zero tone. Then, for example, sol is the 7th tone, la the 9th, and so on. The 12th tone is again do, only an octave higher.

In a tempered chromatic scale, each tone has $\sqrt[12]{2}$ more vibrations than the preceding tone. That means that the number of vibrations of arty tone can be expressed by the formula

$$N_{pm} = n \cdot 2^m (\sqrt[12]{2})^p.$$

"Taking logarithms, we get

$$\log N_{pm} = \log n + m \log 2 + p \frac{\log 2}{12}$$

Or

$$\log N_{pm} = \log n + \left(m + \frac{p}{12}\right) \log 2.$$

Now, taking the number of vibrations of the lowest do as unity (n = 1) and converting all logarithms to base 2 (or simply taking log 2 = 1), we have

$$\log N_{pm} = m + \frac{p}{12}.$$

"From this we see that the numbers of the keys of a piano represent the logarithms of the numbers of vibrations of the appropriate sounds (multiplied by 12). We my even say that the number of the octave is the characteristic, and the number of the sound in the given octave (divided by 12) is the mantissa of that logarithm."

Let us take an example. In the tone sol of the third octave, that is, in the number 3 + 7/12 (≈ 3.583), the number 3 is the characteristic of the logarithm of the number of vibrations of the tone, and 7/12 (≈ 0.583) is the mantissa of that logarithm to the base 2; hence the number of vibrations is $2^{3,583}$ or 11.98 times greater than the number of vibrations of the tone do of the first octave.

The Stars Noise anti Logarithms

This heading with its combination of apparently un-combinable items is not an attempt to parody Kuzma Prutkov and his writings; it does indeed concern stars and noise and logarithms, all closely related.

Noise and stars are grouper together because the loudness of noise and the brightness of stars are both gauged in the same manner, by a logarithmic scale.

Astronomers divide the star according to apparent brightness into first-magnitude stars; second-magnitude, and so on. The sequence of stellar magnitudes is perceived by the human eye as terms in an arithmetic progression. However, the physical brightness varies according to a different law: the objective brightnesses of the stars form a geometric progression with ratio 2.5. It is easy to see that the magnitude of a star is nothing other than the logarithm of its physical brightness. For example, star H of third magnitude are brighter than those of first magnitude by a factor of 2.5^{3-1}, or 6.25. In short, the astronomer estimates the apparent brightness of stars by operating with a table of logarithms to the base 2.5. I now leave this topic because it has been dealt with in sufficient detail in my book entitled *Recreational Astronomy* [in Russian].

The loudness of sound is described in similar fashion. The harmful effect of industrial noises on the health of workers and on the productivity of labor was an impetus to work out ways of an exact numerical evaluation of the loudness of sound. The unit of loudness is the "bel" and the practical unit is a tenth of a bel, 0: the decibel. Successive degrees of loudness -1 bel, 2 bels and so on (practically speaking, 10 decibels, 20 decibels, and so on)-constitute to our ear an arithmetic progression. But the physical intensity of these noises form a geometric progression with common ratio 10. To a loudness difference of 1 bel there corresponds a difference of 10 in the intensity of the noises. This means the loudness of sound expressed in bets is equal to the common logarithm of its physical intensity.

A few examples will hip to clarify this matter.

The soft rustling of leaves is estimated at 1 bel, a loud conversation is put at 6.5 bels, the growl of a lion at 8.7 bels. From this it follows that the sound intensity of a conversation exceeds the rustling of leaves by a factor of

$$10^{6.5-1} = 10^{5.5} = 316\ 000.$$

The growling of a lion is louder than a conversation by a factor of

$$10^{8.7-6.5} = 10^{5.5} = 158.$$

Noise louder than 8 bels is recognized as being injurious to the human organism. Many factories have higher noise levels with noises of 10 and more bels. A hammer blow on a steel sheet generates 11 bels of sound. Such noises are 100 and 1000 times stronger than the permissible level and are 10 to 100 times louder that the loudest spot near the Niagara falls (which is 9 bels).

Is it by accident that in measuring the apparent brightness of stars and the loudness of sound we have to do with a logarithmic relationship between the magnitude of the perception and the generating stimulus? No, both are a consequence of a general law, called Fechner's law which is a psychophysical law that states that the intensity of the sensory response is proportional to the logarithm of the stimulus intensity.

So you see logarithms have their way into psychology too.

Logarithms in Electric Lighting

PROBLEM

The reason why gas-filled lamps produce a brighter light than electric-filament vacuum lamps of the same material lies in the different temperature of the filament. By a rule that has been established in physics, the total amount of light emitted at white incandescence increases with the 12th power of the absolute temperature. With these facts, let us now calculate how many times more light is emitted by a gas- filled lamp whose filament has a temperature of 2500° on the absolute scale (that is, reckoning from -273 Celsius) than by a vacuum lamp with a filament at 2200°.

SOLUTION

Denoting the desired relation by x, we have the equation

$$x = (2500/2200)^{12} = (25/22)^{12}$$

whence

$$\log x = 12 (\log 25 - \log 22), x = 4.6.$$

A gas-filled lamp emits 4.6 times more light than a vacuum lamp. Thus, if a vacuum lamp is rated at 50 watts, then the gas-filled lamp will yield 230 watts under the same conditions.

Let us calculate further to find out what increase in absolute temperature (in per cent) is necessary to double the brightness of the lamp.

SOLUTION

We set up the equation

$$(1 + x/100)^{12} = 2$$

and find that

$$\log (1 + x/100) = \log 2/12 \text{ and } x = 6\%$$

Finally, a third calculation. What is the percentage increase in the brightness of a lamp if the (absolute) temperature of its filament increases by 1%?

SOLUTION

Using logarithms to calculate the following expression,

$$x = 1.01^{12}$$

we find that

$$x = 1.13.$$

The brightness increases 13%.

Calculating for a two per cent increase in temperature, we find a 27 per cent increase in brightness; if the temperature increases 3%, the rightness will increase 43%.

It is now clear why so much attention in the manufacture of electric light bulbs is paid to increasing the temperature of the filament, with every extra degree at a premium.

Making out a Will for Hundreds of Years

Who hasn't heard of the legendary number of grains of wheat that the inventor of chess asked as a reward? That number was built up out of a successive doubling of unity: One grain was asked for the first square on the chessboard, two for the second, and so on, doubling each time until the 64th square was reached.

It will be found however that numbers tend to grow unexpectedly fast not only in the case of successive doubling but even when the rate of increase is rather moderate. Capital invested at 5% interest increases annually by a factor of 1.05. This would not seem to be much of an increase, yet if the time interval is long enough the capital builds up into a tremendous sum. This explains the amazing increase of capital bequeathed for very long periods of time. It seems very strange indeed that a testator can leave a small sum of money and also instructions for the payment of enormous sums. Yet here is the case of the famous United States statesman Benjamin Franklin who left a will of extreme interest. In rough outline it amounts to this.

One thousand pounds sterling is bequeathed by Franklin to the residents of the city of Boston. It is requested that the most illustrious citizens of the city be entrusted with that .sum, which is to be lent at 5 per cent interest annually to young handicraftsmen (there were no institutions of commercial credit in the United States in those days). In one hundred years this sum would increase to 131 000 pounds sterling. He then instructed 100 000 pounds to be used for the construction of municipal buildings and the remaining 31 000 pounds to be invested at interest for 100 years. At the end of the second century, the sum was to have increased to 4 060 000 pounds sterling, of which 1 060 000 pounds were to be left to the residents of Boston to be used at their discretion, while 3 000 000 poulds were to go to the management of the community of Massachusetts. Beyond that, Benjamin Franklin did not risk further uses of his accumulated money.

He left only 1000 pounds but with instructions involving millions. There is no contradiction here, however. A mathematical calculation will show that the reasoning behind this operation is quite realistic. Every year 1000 pounds increase by a factor of 1.05 and in 100 years become $x = 1000\ 1.05^{100}$ pounds.

This expression can be calculated with the aid of logarithms:

$$\log x = \log 1000 + 100 \log 1.05 = 5.11893,$$

which yields

$$x = 131\ 000,$$

in complete agreement with the text of Franklin's will. Then we have 31 000 pounds which during the next century become $y = 31\ 000.1.05^{100}$, whence, using logarithms, we get $y = 4\ 076\ 500$, which is practically the same as that indicated in the will.

I leave it to the reader to tackle the following problem taken from The Messieurs Golovlev of the Russian writer Saltykov-Shchedrin.

"Porfiry Vladimirovich k seated in his study numbering extensively on some sheets of paper. The question that interests him is: How much money would he now have if his dear mother had not taken the 100 rubles given to him by his grandfather when he was born, but had banked it in the name of the young Porfiry? It turned out to be very little, however: only eight hundred rubles."

Assuming that at the time of the calculations Porfiry was 50 and again assuming that he carried out the calculations correctly (which is really hard to believe since Golovlev most likely did not know about logarithms and could not be expected to handle compound interest calculations), it is required to find out the interest he would have received.

Constant Growth of Capital

In savings banks, the interest is added to the principal annually. If the interest is added more frequently, then the capital grows faster because a larger amount of money participates in the formation of interest. Let us take a purely theoretical and extremely simplified case. Suppose a deposit of 100 rubles is made in a savings bank at 100% annual interest. If the interest is added to the principal only at the end of one year, then by that time the 100 rubles becomes 2006 Now let us see what happens if the interest is added to the

$$100 \text{ rubles } 1.5 = 150 \text{ rubles.}$$

After the next half-year period we have

$$150 \text{ rubles } 1.5 = 225 \text{ rubles.}$$

If the interest is added every 1/3 year, then at the expiration of one year the

100 rubles turns into

$$100 \text{ rubles } (11/3)^3 \approx 237 \text{ rubes and } 03 \text{ kopecks}$$

Let us now speed up the adding of interest: we will add it to the capital, say, at intervals of 0.1 year, then 0.01 year, 0.001 year and so on. Then 100 rubles will generate the following sums after one year:

$$100 \text{ rubles } 1.1^{10} \approx 259 \text{ rubles } 37 \text{ kopecks,}$$

$$100 \text{ rubles } 1.01^{100} \approx 270 \text{ rubles } 48 \text{ kopecks,}$$

$$100 \text{ rubles } 1.001^{1000} \approx 271 \text{ rubles } 69 \text{ kopecks.}$$

Higher mathematics can be used to prove that if the time intervals are reduced without limit, the built-up capital does not increase without bound, but rather approaches a certain limit which is approximately[3] equal to

271 rubles and 83 kopecks.

Capital deposited at 100% can never increase faster than 2.7183 times even if the interest is added to the principal every second.

The Number *e*

The number 2.718... plays a fundamental role in higher mathematics (probably not less significant than the famous number a) and has a special symbol: e. This number is irrational, which means it cannot be exactly expressed by any finite number of digits[4] and is computed in approximate fashion only -to any desired degree of accuracy- by the following series:

$$1 + 1/1 + 1/1.2 + 1/1.2.3 + 1/1.2.3.4 + 1/1.2.3.4.5 + \dots$$

From the example given above about the growth of capital in terms of compound interest, it is easy to see that the number e is the limit of the expression

$$(1 + 1/n)^n$$

as n increases without bound.

-3- Fractions of kopecks were dropped.
-4- Also, this number, like π, is transcendental, which means it cannot he obtained by solving any algebraic equation involving integral coefficient

For many reasons which we cannot go into here the number e is highly desirable as a base for logarithms. Such tables (tables of "natural logarithms") exist and are extensively used in science and engineering. The champion logarithms involving 48, 61, 102 and 260 digits that we spoke of a little while ago use the number e for their base.

The number e often puts in an appearance where it is least of all expected. Let us take a look at the following problem.

How should one partition a given number a so that the product of all its parts is a maximum?

We already know that the largest product for a constant sum is obtained when the numbers are all equal. Clearly, the number a is to be partitioned into equal parts. But into how many equal parts? Two, three or ten? Techniques in higher mathematics enable us to establish that the largest product is obtained when the parts are as close as possible to e.

For example, partition 10 into a number of equal parts such that they are as close as possible to 2.718.... To do this we have to find the quotient

$$10/2.718\ldots = 3.678\ldots.$$

Since it is not possible to partition the number into 3.678... equal parts, we choose the closest whole number 4 as the divisor. Thus, we obtain the largest product of the parts of 10 if the parts are equal to 10/4 or 2.5.

And so

$$(2.5)^4 = 39.0625$$

is the largest number that can be obtained from multiplying together equal parts of the number 10. Indeed, dividing 10 into 3 or 5 equal parts, we get smaller products:

$$(10/3)^3 = 37,$$

$$(10/5)^5 = 32.$$

In order to obtain the largest product of the parts of 20, the number has to be partitioned into 7 equal parts because

$$20 : 2.718\ldots = 7.36 \approx 7.$$

The number 50 has to be partitioned into 18 parts and the number 100 into 37 parts because

$$50 : 2.718\ldots = 18.4$$

$$100 : 2.718\ldots = 36.8$$

The number e plays a tremendous role in mathematics, physics, astronomy and other sciences. Here are some of the questions considered mathematically that involve e (the list could be extended indefinitely):

Barometric height formula (decreasing pressure with increasing height),

Euler's formula (see the second part of my Physics for Entertainment: the chapter entitled "Jules Verne's Strong Man and Euler's Formula"),

The law of cooling of bodies,

Radioactive decay and the age of the earth,

Oscillations of a pendulum in the air,

Tsiolkovsky's formula for rocket speeds (see my book *Interplanetary Travel*),

Oscillatory phenomena in a radio circuit,

The growth of cells.

Logarithmic Comedy

PROBLEM

Here is another one of those mathematical comedies played out in Chapter 5: prove that 2 > 3. This time we make use of logarithms. The comedy starts out with the inequality

$$1/4 > 1/8$$

which is definitely correct. We then transform to

$$(1/4)^2 > (1/2)^3$$

which is unquestionably clear. To the greater number there corresponds the greater logarithm, and so

$$2\log_{10}(1/2) > 3\log_{10}(1/2)$$

Cancelling out log10 (1/2) we are left with: 2 > 3. What is wrong with this proof?

SOLUTION

The trouble is that when we cancelled out $\log_{10}(1/2)$ we forgot to reverse the sign of the inequality (> to <): yet this was necessary because $\log_{10}(1/2)$ is a negative number.

If we had taken logs to a base n of less than 1/2 instead of 10, then $\log_n(1/2)$ would be positive, but then we couldn't have asserted that the greater number is associated with the larger logarithm.

Any Number via Three Twos

PROBLEM

And now we end this book with a witty algebraic brainteaser that amused the Participants of a congress of physicists in Odessa. The problem is to represent any number that must be positive and whole (any positive integer) using three twos and mathematical symbols.

SOLUTION

Let us take a particular case. Suppose we are given the number 3. Then the problem is solved thus: $3 = -\log_2 \log_2 \sqrt{\sqrt{\sqrt{2}}}$

It is easy to see that this equation is true. Indeed,

$$\sqrt{\sqrt{\sqrt{2}}} = \left[\left(2^{\frac{1}{2}}\right)^{\frac{1}{2}}\right]^{\frac{1}{2}} = 2^{\frac{1}{2^3}} = 2^{2^{-3}},$$

$$\log_2 2^{2^{-3}} = 2^{-3}, \quad -\log_2 2^{-3} = 3.$$

If we were given the number 5, we would proceed in the same manner:

$$5 = -\log_2 \log_2 \sqrt{\sqrt{\sqrt{\sqrt{\sqrt{2}}}}}.$$

It will be seen that we have made use of the fact that the index 2 is dropped when writing the square root. The general solution looks like this. If the given number is N, then

$$N = -\log_2 \log_2 \underbrace{\sqrt{\sqrt{\ldots\sqrt{\sqrt{2}}}}}_{N \text{ times}},$$

the number of radical signs equaling the number of units in the given number.

ISBN : 9782917260265